Structural Integrity and Failure

Edited by Resat Oyguc and Faham Tahmasebinia

Published in London, United Kingdom

IntechOpen

Supporting open minds since 2005

Structural Integrity and Failure
http://dx.doi.org/10.5772/intechopen.87855
Edited by Resat Oyguc and Faham Tahmasebinia

Contributors
Khawaja Ali, Aleena Saleem, Ganga Kv Prakhya, Ruodan Lu, Chris Rausch, Marzia Bolpagni, Ioannis Brilakis, Carl T. Haas, Louay Yousuf, M. Rosário Oliveira, Mersida Manjgo, Meri Burzic, Konstantinos Koulouris, Charis Apostolopoulos, Resat Oyguc, Evrim Oyguc, Abdul Hayır

First published in London, United Kingdom, 2021 by IntechOpen
IntechOpen is the global imprint of INTECHOPEN LIMITED, registered in England and Wales, registration number: 11086078, 5 Princes Gate Court, London, SW7 2QJ, United Kingdom
Printed in Croatia

British Library Cataloguing-in-Publication Data
A catalogue record for this book is available from the British Library

Additional hard and PDF copies can be obtained from orders@intechopen.com

Structural Integrity and Failure
Edited by Resat Oyguc and Faham Tahmasebinia
p. cm.
Print ISBN 978-1-83881-980-4
Online ISBN 978-1-83881-981-1
eBook (PDF) ISBN 978-1-83881-982-8

For EU product safety concerns:
IN TECH d.o.o., Prolaz Marije Krucifikse Kozulić 3, 51000 Rijeka, Croatia, info@intechopen.com or visit our website at intechopen.com.

Meet the editors

Dr. Resat Oyguc is an Associate Professor in the Department of Earthquake Engineering at the Institute of Disaster Management at Istanbul Technical University. He received his M.Sc. and Ph.D. degrees from Yıldız Technical University in Structural Engineering, and Istanbul Technical University in Structural and Earthquake Engineering, respectively. His research doctorate was on the seismic capacity assessment of irregular structures using adaptive pushover procedures. He investigates the response of complex networks and structures under extreme loading conditions. He has also undertaken many studies on the safety assessment of structures and evaluation of post-earthquake failures. Specifically, he works on nonlinear inelastic dynamic response simulation for infrastructures. His recent research includes seismic performance assessment of a nuclear reactor building under multiple earthquake excitations considering the degradation effects.

Faham Tahmasebinia holds ME and ME-Research degrees in Civil/Structural Engineering from the University of Wollongong – Australia. He has also completed two Ph.D. degrees in the field of Structural Engineering at the University of Sydney and in the field of Rock Mechanics at the University of New South Wales – Sydney. Currently, he is an academic at the University of Sydney – Australia. His research areas are numerical and analytical simulations in both ductile and brittle materials.

Contents

Preface

Structural integrity and failure assessment have been considered by many fields of engineers in the literature as it is a multi-disciplinary concept. The assessment procedure vitally ensures that structural elements will remain functional throughout their service lives. Structural failure refers to the loss of structural integrity by means of loss at the component- or system-level elements. The main concern of integrity assessment as an aspect of engineering is that a structural failure may be avoided at the service level by designing the structure to withstand its designated loads. Hence, for satisfactory structural performance, structural safety, failure, and interaction between them should be taken into account throughout the design and analysis stages.

This book is a collection of chapters to provide the researcher with a comprehensive perspective on structural integrity and its sub-disciplines. Topics of interest include: structural integrity, sustainable structural design and analysis, failure analysis and case studies, damage, fatigue, and fracture of materials and structures, durability, safety and reliability analysis of structural components, and structural health monitoring.

The book is organized into two sections: Structural Integrity and Failure; and Bridge Engineering. Both sections include four chapters. In the first section, Chapter I discusses a procedure for monitoring fatigue crack growth. Chapter II assesses the seismic behavior of a sample third-generation nuclear reactor building and observes the stress distributions and crack propagations of the prestressed outer containment. Chapter III covers the effects of steel corrosion on bond relationship between steel and concrete. Chapter IV presents a methodology for developing control, measurement and monitoring plans. In the second section, Chapter V studies the derivation of analytic formulation of bending deflection using the theory of classical laminate plate. Chapter VI focuses on the temporary works methods applied to enable the construction of bridges near third party assets by means of independent checks, monitoring, and back analysis. Chapter VII presents an exploratory analysis of the geometric accuracy of digital twins generated for existing infrastructure using point clouds. Chapter VIII describes the evaluation of safety factors of stay cables by employing the deterministic and non-deterministic methods at limit states.

The editor thanks the authors of the chapters and all contributors. Without their participation this book would not have been possible. Further, the efforts of the staff at IntechOpen in bringing this book to fruition are highly appreciated.

Dr. Resat A. Oyguc
Associate Professor,
Istanbul Technical University,
Turkey

Faham Tahmasebinia
The University of Sydney,
Australia

Section 1

Structural Integrity and Failure

Chapter 1

Treatment Analysis of Welding Structure in the Presence of a Crack Type Defects

Mersida Manjgo and Meri Burzic

Abstract

The largest number of welded structures in operating conditions is exposed to variable loads, which is why the share of fatigue fracture in the failure of welded structures is higher than others. The essence of construction with fracture safety is that the structure can withstand the designed load in the designed time. If a crack is detected during operation, it is possible to predict the development of damage during the service life as well as the load-bearing capacity of the structure depending on the development of damage. The paper describes a new system for monitoring fatigue crack growth, which is based on the change in the resistance of the measuring foil during crack growth. The system is compatible with the basic settings of the ASTM E647–86 standard, which refers to the determination of the fatigue crack growth rate.

Keywords: welded joint, fatigue crack, fatigue threshold, crack growth rate

1. Introduction

Mass application of welded structures began with the development of welding procedures on the one hand and the development of steels with suitable properties on the other. Along with welding processes, in parallel, methods for assessing the safety of welded joints were also developed.

Construction materials and welded joints can contain defects and microcracks that are the beginnings of fractures. Exploitation conditions can lead to cracking even if there are no defects in the material, e.g., at places of stress concentration caused by the design of the structure. Under the influence of unfavorable exploitation factors, such as fatigue and corrosion, cracks can grow steadily, and after enough time, reach a critical size and cause breakage.

Fatigue is the phenomenon of gradual destruction of a material due to the long-term action of a periodically changing load. Damage to structures, caused by material fatigue, represents 50 ÷ 90% of all damage to structures in exploitation [1]. The significance of fatigue damage is obvious because a large number of such damages lead to catastrophic fractures.

The traditional, well-known, S-N approach is based on the experimental determination of the dependence of the stress amplitude from the number of cycles to fracture.

This standard method is built into many standards and regulations and is widely used in the design of welded and other structures. In this test, as a rule, only the number of changes of the load to fracture under the action of a constant range load is determined, and the standard only requires information on the magnitude of the stress at which crack and fracture initiation does not occur after a certain number of cycles (usually between 10^6 and 10^8 cycles).

In the presence of cracks, the question arises of its development under the action of a variable load.

Fatigue crack growth is a very complex process that depends on a number of variables [2]:

- the intensity of the effective stress field at the crack tip defined by the K-factor;
- type and form of load;
- environment (aggressiveness, temperature, humidity),
- mechanical and metallurgical characteristics of the material

The method of fracture mechanics is based on linearly elastic fracture mechanics and originates from the Paris' law of 1962, and is still applied, although the impact of large plastic deformations around the crack tip has not been fully taken into account. The constants that occur such as "C" i "m" in Paris' law $da/dN=C\ \Delta K^m$, must be determined separately for each material and the specific test conditions. These data are essential for three types of fatigue analysis:

- to accurately determine fatigue crack behavior
- to estimate the life of the structure
- to calculate fatigue damage

2. Determination of dynamic characteristics of welded joint

Metal fatigue is defined as the process of cumulative damage under the action of variable load, which is manifested by the appearance of fatigue cracks and fractures. The fatigue strength of welded joints is determined by testing the specimens at variable load until a crack or fracture occurs.

The test was performed on a high-frequency AMSLER pulsator. The high-frequency pulsator can achieve a sinusoidal alternating load in the range from −100 kN to +100 kN. In order to more fully assess the behavior of the material under the action of variable load, and having in mind the dimensions of the specimen, the most critical case of the action of variable load was made, namely alternating variable load tension - pressure ($R = -1$), **Figure 1**.

It is clear that the strength at high cyclic fatigue depends on the properties of the constituents of the welded joint. Therefore, data are needed for BM and WM, but also for HAZ, which makes testing of welded joints in high-cyclic fatigue complex and expensive. The aim of the test is to determine the points in the S-N diagram (construction of the Wehler curve) and to determine the permanent dynamic strength S_f. The test procedure as well as the specimen are defined according to ASTM E466 [3]. The appearance of the test tube with variable load is shown in **Figure 2**.

The determination of the maximum dynamic stress at which no crack-type error is initiated in smooth construction forms is shown graphically in the form of Weller curves (S-N diagrams) in **Figure 3**. for butt-welded joint specimen and **Figure 4** for specimen removed from BM.

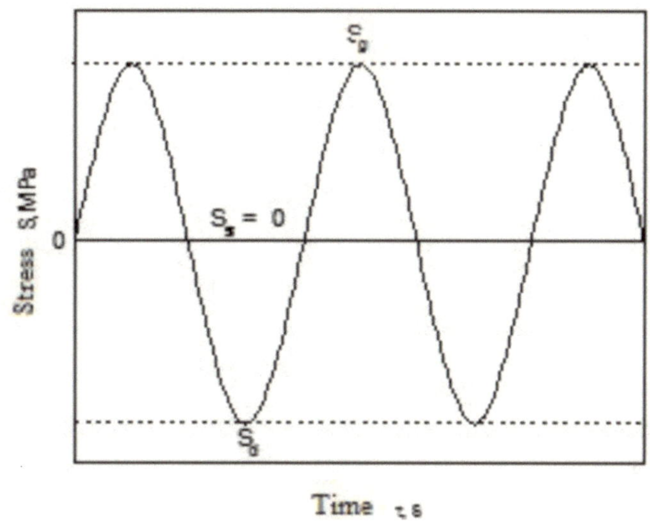

Figure 1.
Alternating load scheme. R = −1.

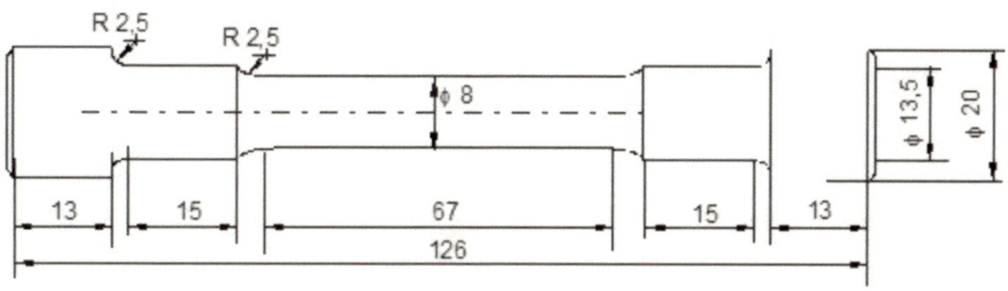

Figure 2.
Dynamic specimen according to ASTM E466.

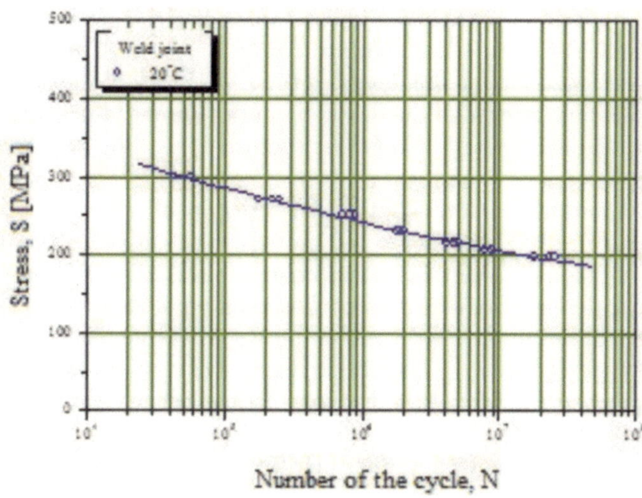

Figure 3.
S-N diagram of speciemn taken out of butt welded joint and tested at room temperature.

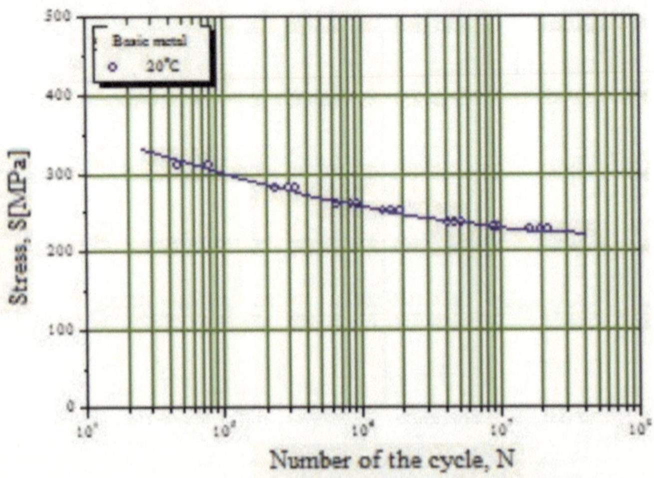

Figure 4.
S-N diagram of speciemn extracted from the new BM and tested at room and operating temperatures.

To construct one Weller curve and determine the permanent dynamic strength, it is necessary to test the specimen at 6 to 7 different load levels. According to the ASTM E 466 standard, three specimen were tested for each load level, which is a total of 21 specimen. Therefore, this test is extremely expensive and justified when design data are required, primarily from the aspect of fatigue and fracture mechanics. So when designing parts exposed to long-term variable load in the total design life of the structure.

The traditional, well-known, S-N approach is based on the experimental determination of the dependence of the stress amplitude from the number of cycles to fracture.

3. Fatigue analysis from fracture mechanics angle

The origin of the fault may be related to design and construction, technology, and producing the structure, control, and testing. The continuance of construction then depends only on the possibility and conditions of crack growth from the initial failure. The essence of construction with fracture safety is that the structure can withstand the designed load in the designed time. If a crack is detected during operation, it's possible to predict the development of damage during the service life as well as the load-bearing capacity of the structure depending on the development of damage [4].

The most important characteristics for the operational safety of structures are the ones that describe the appearance and growth of cracks under the influence of variable load. A generally accepted characteristic, in this case, is fatigue strength. Accordingly, the design of structural parts based on possible material fatigue is based on the use of fatigue strength, and empirical recommendations, derived from the analysis of parts failure in operation and extensive testing [4].

The appearance of a fatigue crack requires that the behavior of the material around the crack tip is considered based on the micromechanical aspect. The initiation of cracks and their growth condition that the micromechanical aspect of the behavior of the material becomes important for the assessment of the operational safety of structural parts. The existence of a singularity in the form of a crack tip indicates the application of fracture mechanics and its parameters, such as stress intensity factor, crack opening, and contour J-integral. Paris' crack growth law, which determines the dependence of the load and the corresponding range of stress intensity factors, with the crack growth rate per cycle, is generally accepted today [4].

3.1 Fatigue crack growth rate da/dN i ΔK_{th}: Paris' law

The need to introduce fracture mechanics into the study of fatigue behavior arose from the analysis of crack growth under cyclic loading.

Many data on fatigue crack growth were obtained by examining CT specimen, which were exposed to loads with a constant amplitude ΔP, **Figure 5** [5].

In **Figure 5** a typical form of crack length dependence on the number of load cycles *a*-N for three load range levels ΔP = const where $\Delta P_1 < \Delta P_2 < \Delta P_3$ is shown.

It is noticeable that with the increasing number of cycles N and crack length "a", the crack growth rate defined by the slope of the tangent increases steadily. Also, with the increase in the load range ΔP, there is a faster increase of the speed gradient. In other words, the crack, for example, of length a1 in **Figure 1**, grows faster at the load amplitude ΔP_3 than at the load ΔP_2 or ΔP_1.

Numerous theoretically and empirically defined dependencies in the form $da/dN = f(P, a)$ can be found in the literature, which emphasizes the importance of load and cracks length. The first to define the range of stress intensity factor $\Delta K = f(\sigma, a)$ in the form of fatigue crack growth rate as a basic parameter were Paris and co-workers [5].

$$\Delta K = K_{max} - K_{min} = Y(\sigma_{max} - \sigma_{min})(\pi a)^{1/2} = Y\Delta\sigma(\pi a)^{1/2} \tag{1}$$

The crack growth rates *da/dN* as a function of ΔK are determined from the corresponding curve a-N, graphically, or numerically. The experimental results presented on the double-logarithmic scale usually have a characteristic S-shape, schematically shown in **Figure 6**.

It is noticeable that the crack propagation is initially accelerated (area I), then passes into the phase of stable growth (area II), to finally pass into the phase of critical crack expansion (area III).

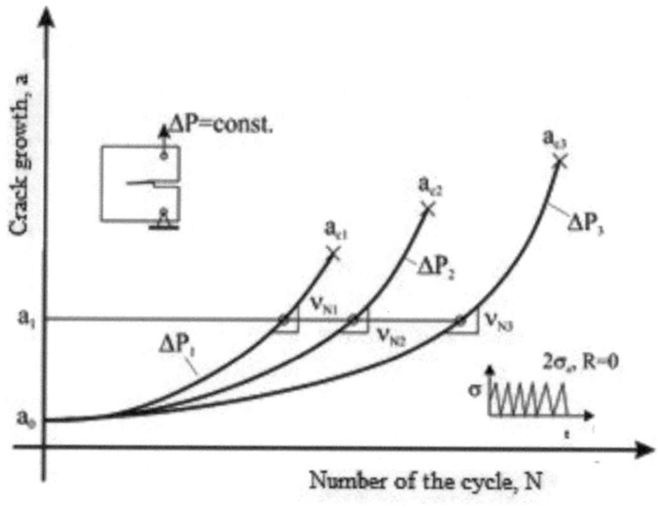

Figure 5.
Crack growth dependence a = f(N) for three levels of load ranges ΔP = const where ΔP1 < ΔP2 < ΔP3.

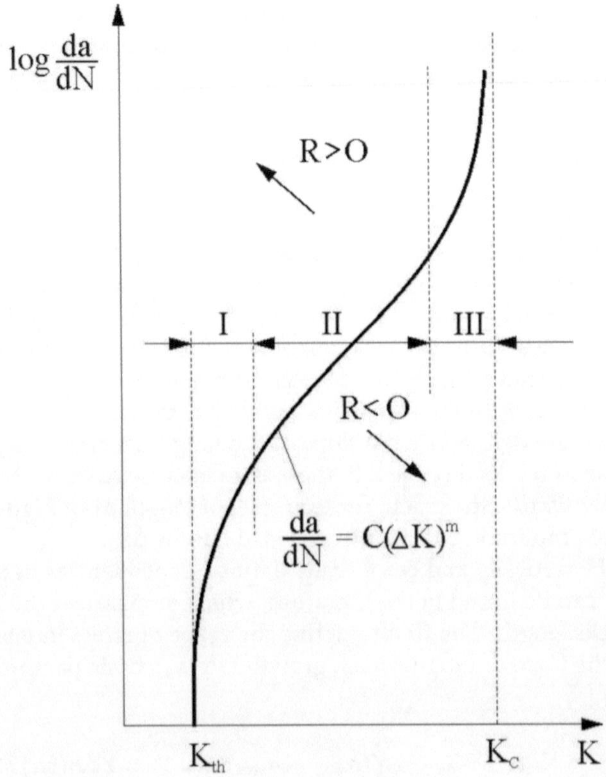

Figure 6.
The principle form of change of growth rate da/dN = f(ΔK) for R = and directions of displacement of S-curve for relations R≠o.

From the point of view of the crack growth mechanism and different intensities of influencing factors, three areas can be observed on this curve [6–8].

In area I, the crack propagation velocity tends to zero as the range of stress intensity factors approaches the crack propagation threshold:

$$\Delta K_{th} = K_{th,\max} - K_{th,\min} = \Delta\sigma\sqrt{\pi \cdot a_{th}} \cdot Y \quad (2)$$

where: a_{th}- the length of the initial crack.

In area II the crack grows linearly in the log–log diagram, so it can be described by the equation:

$$\frac{da}{dN} = C \cdot \Delta K^m \quad (3)$$

where C and m are the material constants determined experimentally.

This law is known as Paris' law.

In area III, the crack grows rapidly as the range of stress intensity factors approaches ΔK_c:

$$\Delta K_c = K_{Ic} - K_{c,\min} = \Delta\sigma\sqrt{\pi \cdot a_c} \cdot Y \quad (4)$$

where a_c - critical crack length, K_{Ic} - fracture toughness.

As Paris' law is valid only in area II, attempts were made to find equations that would describe crack growth in other areas of growth as well. One such is the Forman equation describing crack growth in areas II and III:

$$\frac{da}{dN} = \frac{C \cdot \Delta K^m}{(1 - R) \cdot K_{Ic} - \Delta K} \tag{5}$$

Klesnil and Lucas modified the Paris law, taking into account the crack propagation threshold, and thus obtained the crack growth equation valid in areas I and II:

$$\frac{da}{dN} = C \cdot \left(\Delta K^m - \Delta K_{th}^m\right) \tag{6}$$

McEvily developed an expression that is valid for the whole crack growth curve, and which, unlike the previous equations, which were obtained empirically, is based on a simple physical model:

$$\frac{da}{dN} = C \cdot (\Delta K - \Delta K_{th})^2 \left(1 + \frac{\Delta K}{K_{Ic} - K_{\max}}\right) \tag{7}$$

From all these Eqs. (4), (5), (6) and (7) integration can obtain the time required for crack growth from any arbitrary to a critical length. Also, all the above terms are valid in the case of type i loads.

3.2 Determination of fatigue crack growth parameters

The basic progress that fracture mechanics has made in the sphere of material fatigue is in the analytical breakdown of the fatigue fracture phenomenon into the period of creation, in which the fatigue crack occurs, and the period of growth or expansion that follows and in which the resulting crack increases to a critical size at which a sudden fracture occurs. Thus, the total number of cycles, N_u, after which a fracture occurs, is divided by the number of cycles required for the fatigue crack to form, N_i, and the number of cycles for it to increase to the critical value for fracture, N_p, i.e., $N_u = N_i + N_p$, **Figure** 7 [5].

Analysis of the stress state and deformation at the top of a rising fatigue crack by linear elastic fracture mechanics (LEFM) led to the formulation of the Paris equation for all metals and alloys, which relates the fatigue crack growth rate to the stress intensity range at the crack tip:

Although the Paris cracks growth equation is not valid in the whole range, between low velocities near the fatigue threshold ΔK_{th}, and high velocities K_{Ic}, the large linear midpoint of the curve covered by the Paris relation proved to be by far the most important from a practical point of view and fatigue crack growth.

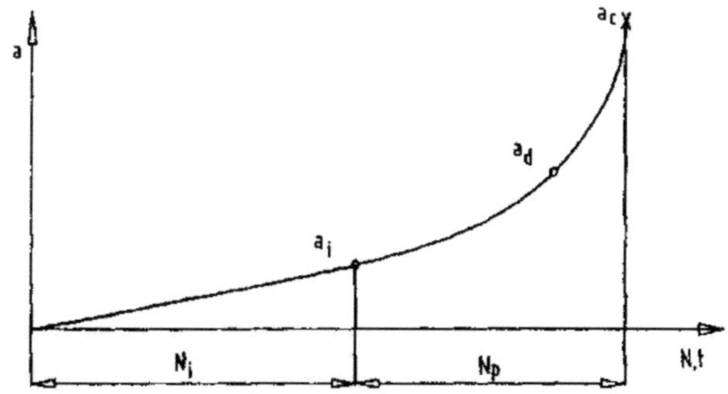

Figure 7.
The share of the initiation period N_i and the fatigue crack growth period N_p in the total fatigue life N_u.

The application of the Paris equation has proved particularly fruitful in the field of fatigue of structures made of high and very-high strength materials.

The ASTM E647 standard [9] prescribes the measurement of the fatigue crack growth rate da/dN, which develops from an existing crack, and the calculation of the stress intensity factor range, ΔK. This means that the test tube should have a tiring crack. There are two important limitations in the ASTM E647 standard: the growth rate must be greater than 10–8 m/cycle to avoid the fatigue threshold area, ΔKth, and the load should be of constant amplitude [4].

Steel of quality A-387 Gr was used to determine the dependence of the fatigue crack growth rate per cycle da/dN and the range of stress intensity factors ΔK. 91 15 mm thick [10]. The chemical composition and mechanical properties of the base material are given in **Tables 1** and **2**.

The welded joint is made with two welding processes and two additional materials.

- Root welding - TIG welding, additional material is wire marked BOEHLER C 9 MV-IG, diameter 2,4 mm (international designation W CrMo 91 according to EN ISO 21952-A).

- Filling - REL welding, additional material is an electrode marked BOEHLER FOX C9 MV, diameter 3,50 and 4,00 mm (international designation E CrMo 91 B 4 2 H5 according to EN ISO 3580-A).

Determination of fatigue crack growth rate da/dN and fatigue threshold ΔK_{th} was performed on standard Charpy tubes by the method of bending the tube at three points on a resonant high-frequency pulsator.

The test was performed at the same minimum and maximum load ratio R = −1. The achieved frequency ranged from 175 to 195 Hz depending on whether the crack passed through the base metal, weld metal, or the heat-affected zone, and on the magnitude of the load. The medium load and its amplitude were registered with accuracy ±3 Ncm.

Prior to the test, the specimen were mechanically prepared and measuring tapes - foils were glued to the prepared tubes, with which the crack growth was monitored. RMF A-5 measuring foils with a measuring length of 5 mm were used for testing. In order to be able to monitor the crack growth using a measuring foil, the FRACTOMAT crack growth detection device was used, **Figure 8**.

Chemical composition, mas. %										
C	**Si**	**Mn**	**P**	**S**	**Cr**	**Mo**	**Ni**	**V**	**Nb**	**Cu**
0.129	0.277	0.443	0.001	0.001	8.25	0.874	0.01	0.198	0.056	0.068

Table 1.
Chemical composition of the tested batch of steel SA 387 Gr. 91.

Yield strength R_{p02} [MPa]	Tensile strength R_m [MPa]	Elongation A [%]	Impact energy Kv [J] +20°C
445	580–760	18	40

Table 2.
Mechanical properties of the base material.

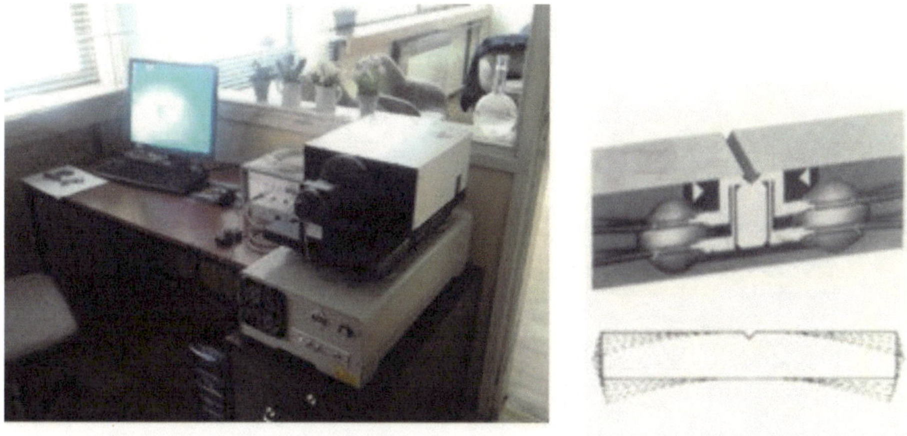

Figure 8.
Modern system for dynamic tests [10].

The scheme of the measuring foil and the method of registering crack growth is shown in **Figure 9**.

The appearance of the prepared specimen for determining the fatigue crack growth parameters is given in **Figure 10**.

Determining the dependence of the fatigue crack growth rate on the cycle da/dN and the range of stress intensity factors ΔK is reduced to determining the coefficient **C** and the exponent **m** in the Paris equation. The fatigue crack growth rate should be attributed to the current crack length, and, to the range of stress intensity factor, ΔK, which depends on the specimen geometry and crack length, and to the variable force range, $\Delta F = F_g - F_d$.

Determining the stress intensity factor range uses the formula

$$\Delta K = \frac{\Delta F \cdot L}{B\sqrt{W^3}} \cdot f(a/W) \tag{8}$$

where: L - range of supports, mm;
B - specimen thickness, mm;
W - width (height) of the specimen, mm, and.
a - crack length.

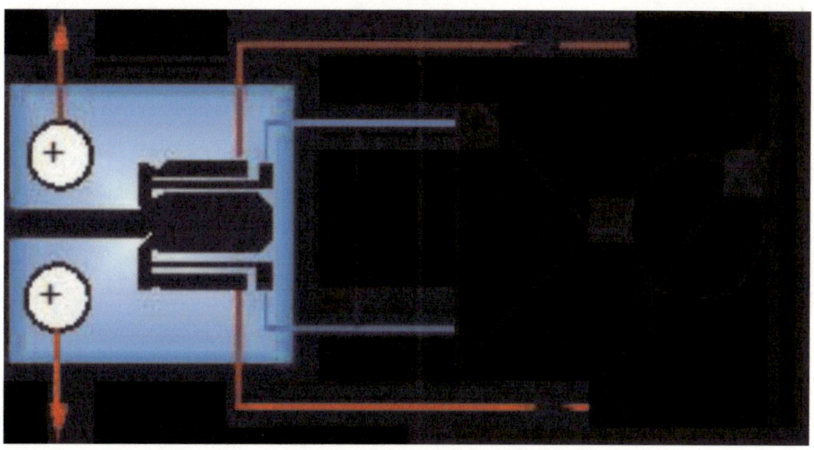

Figure 9.
The scheme of the measuring foil and the method of registering crack growth.

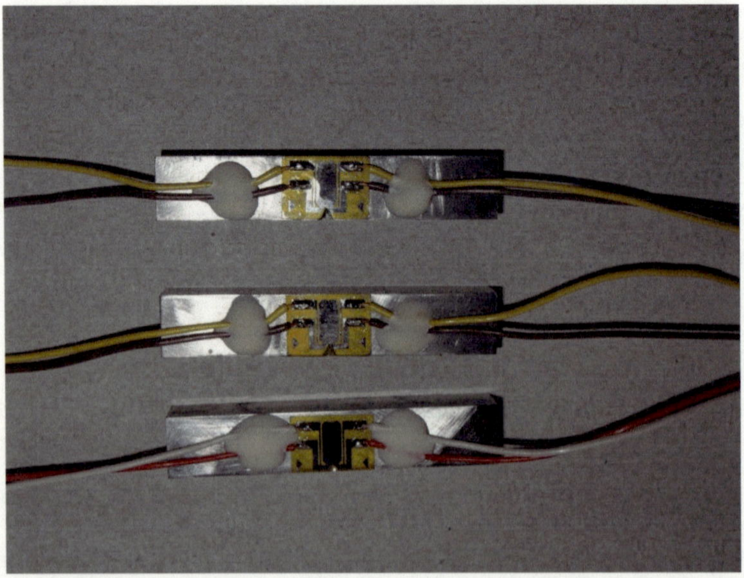

Figure 10.
The appearance of the prepared specimen for parameter testing [10].

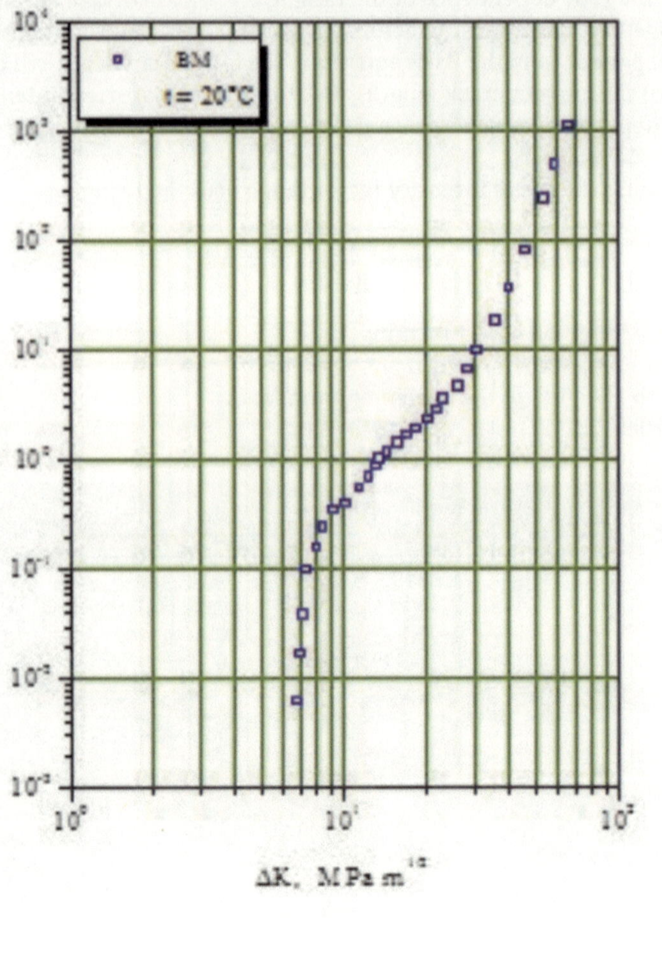

Figure 11.
Dependence diagram da/dN - ΔK for specimens with fatigue crack tip in BM [10].

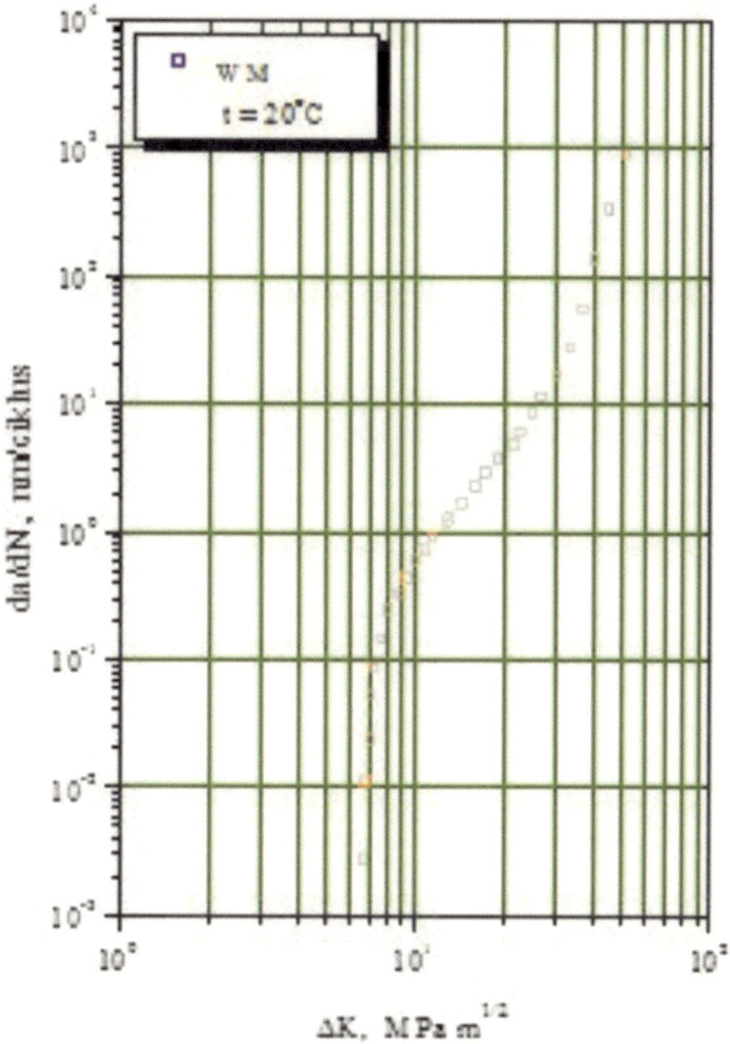

Figure 12.
Dependence diagram da/dN - ΔK for speciemens with fatigue crack tip in WM [10].

The geometric term f (a/W) is given by the expression:

$$f(a/W) = \frac{3 \cdot \sqrt{\frac{a}{W}} \cdot \left[1,99 - \frac{a}{W}\left(1 - \frac{a}{W}\right)\left(2,15 - 3,93\frac{a}{W} + 2,7\left(\frac{a}{W}\right)^2\right)\right]}{2\left(1 + 2\frac{a}{W}\right)\left(1 - \frac{a}{W}\right)^{3/2}} \tag{9}$$

Three groups of specimens depending on the location of the crack tip were examined, namely:

Group I - specimens with a crack tip in BM,

Group II - specimens with a crack tip in WM and.

Group III - specimens with a crack tip in HAZ.

Based on the test flow and the obtained dependences of crack length **a** - number of cycles **N**, the fatigue crack growth rate da/dN is calculated. Depending on the applied variable load expressed through the change of the range of voltage intensity factor, ΔK, log da/dN - log(ΔK) curves are drawn.

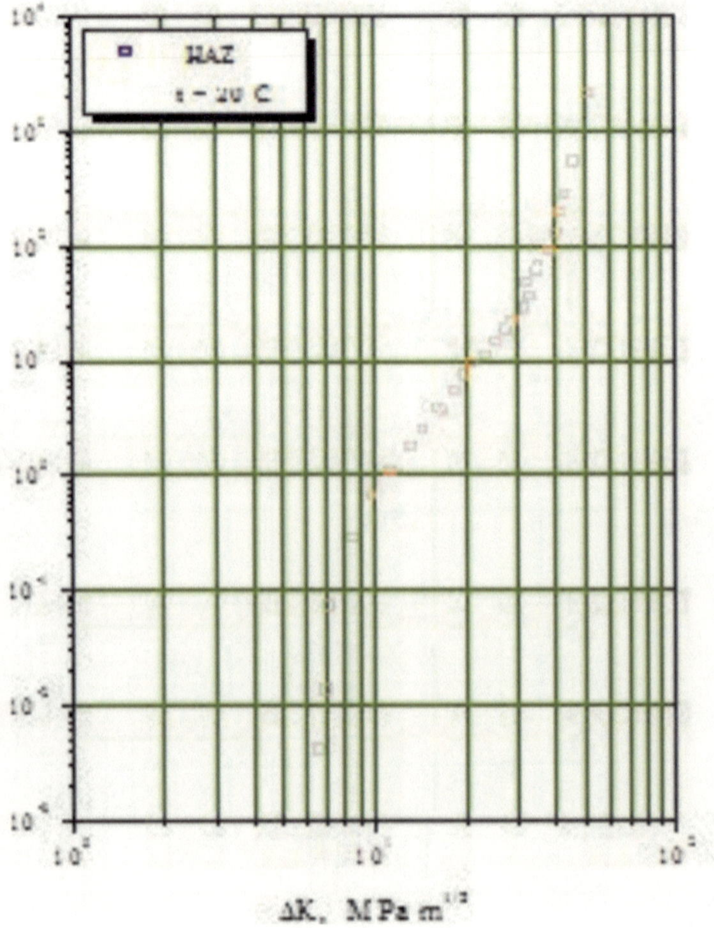

Figure 13.
Dependence diagram da/dN - Δ for specimens with a fatigue crack tip in HAZ [10].

Characteristic diagrams of fatigue crack growth rates, da/dN - change of stress intensity factor range, ΔK, for specimens with fatigue crack tip in BM, **Figure 11.**, for specimens with fatigue crack tip in WM, **Figure 12.**, and for specimens with a fatigue crack tip in HAZ, **Figure 13.**

The obtained values of the parameters of the Paris equation, coefficient C and exponent m, fatigue threshold ΔK_{th}, and fatigue crack growth rate, da/dN, at the value of ΔK = 10 MPa $m^{1/2}$, are given in **Table 3** for specimens with a notch in BM, **Table 4** for specimens with a notch in WM, and **Table 5** for speciemns with a notch in HAZ.

Test tube label	Test temperature, °C	Fatigue threshold ΔK_{th}, MPa $m^{1/2}$	Coefficient C	Exponentm	da/dN, μm/cikl, pri ΔK = 10 MPa $m^{1/2}$
OM-1	20	6,9	$2,98 \cdot 10^{-10}$	3,62	$1,24 \cdot 10^{-6}$
OM-2		6,8	$3,07 \cdot 10^{-10}$	3,58	$1,17 \cdot 10^{-6}$
OM-3		7,1	$2,85 \cdot 10^{-10}$	3,59	$1,11 \cdot 10^{-6}$

Table 3.
Fatigue crack growth parameters for notched specimens in BM.

Test tube label	Test temperature, °C	Fatigue threshold ΔK_{th}, MPa $m^{1/2}$	Coefficient C	Exponentm	da/dN, μm/cikl, pri $\Delta K = 10$ MPa $m^{1/2}$
MŠ-1	20	7,2	$3,88 \cdot 10^{-10}$	3,62	$2,56 \cdot 10^{-6}$
MŠ-2		7,1	$4,05 \cdot 10^{-10}$	3,71	$2,07 \cdot 10^{-6}$
MŠ-3		7,4	$3,93 \cdot 10^{-10}$	3,80	$2,48 \cdot 10^{-6}$

Table 4.
Fatigue crack growth parameters for notched specimens in WM.

Test tube label	Test temperature, °C	Fatigue threshold ΔK_{th}, MPa $m^{1/2}$	Coefficient C	Exponentm	da/dN, m/cikl, pri $\Delta K = 10$ MPa $m^{1/2}$
ZUT-1	20	6,6	$3,05 \cdot 10^{-10}$	4,01	$3,12 \cdot 10^{-6}$
ZUT-2		6,8	$3,07 \cdot 10^{-10}$	4,04	$3,37 \cdot 10^{-6}$
ZUT-3		6,5	$2,85 \cdot 10^{-10}$	4,09	$3,51 \cdot 10^{-6}$

Table 5.
Fatigue crack growth parameters for notched specimens in HAZ.

4. Conclusions

Practice has shown that structures that are exposed to the effects of variable load during operation are most prone to accidents and fractures. The causes are primarily errors that occur in exploitation.

- The reason for the application of fracture mechanics is based on the fact that in the presence of errors, which inevitably occur as a result of imperfections in production processes and/or in operation, there is a loss of load-bearing structure elements and time as a whole.

- The reduction of load-bearing capacity in order to achieve the reliability and safety of structures must be controlled.

- The obtained values of the fatigue threshold ΔKth represent important data on the quality of the tested materials from the aspect of its micromechanical properties, ie on the behavior in the presence of cracks. After its determination, the magnitudes and periods of action of the variable load are known for the formation and then for the growth of the tired crack until cracking.

Author details

Mersida Manjgo[1*] and Meri Burzic[2]

1 Džemal Bijedić University, Faculty of Mechanical Engineering, Mostar, Bosnia and Herzegovina

2 University of Belgrade, Innovation Center, Faculty of Mechanical Engineering, Belgrade, Serbia

*Address all correspondence to: mersida.manjgo@unmo.ba

References

[1] Fuchs H. O., Stephens R. I. Metal Fatigue in Engineering. A Wiley-Intersciene Publication. New York: John Wiley and Sons; 1980.

[2] Podrug S. Mehanika loma. Split: Fakultet elektrotehnike, strojarstva i brodogradnje; 2009.

[3] ASTM E466–82. Standard Practice for Conducting Constant Amplitude Axial Fatigue Test of Metalic Materials. Annual Book of ASTM Standard; 1986.

[4] M. Manjgo, Kriterijumi prihvatljivosti prslina u zavarenim spojevima posuda pod pritiskom od mikrolegiranih čelika, Doktorska disertacija, 2008

[5] V. Ćulafić, Uvod u mehaniku loma, Podgorica

[6] Heckel K. Einfuhrung in die technische Anwendung der Bruchmechanik. Munchen: Carl Hanser Verlag; 1970.

[7] Ekberg A. Fatigue crack propagation. Gothenborg: Chalmers Solid Mechanics; 2003.

[8] Ritchie R. O. Mechanisms of fatigue-crack propagation in ductile and brittle solids. Netherlands: International Journal of Fracture 100: 55–83, 1999.

[9] ASTM 647 Standard Test Method for Constant-Load-Amplitude Fatigue Crack Growth

[10] M. Manjgo, M. Burzić, Optimal welding parameters of SA 387 Gr. 91 thick steel plates in corrosive environment, Elobarat po projektu Eureka, Beograd-Mostar, 2016–2019.

Chapter 2

Structural Modeling and Dynamic Analysis of a Nuclear Reactor Building

Evrim Oyguc, Abdul Hayır and Resat Oyguc

Abstract

Increasing energy demand urge the developing countries to consider different types of energy sources. Owing the fact that the energy production capacity of renewable energy sources is lower than a nuclear power plant, developed countries like US, France, Japan, Russia and China lead to construct nuclear power plants. These countries compensate 80% of their energy need from nuclear power plants. Further, they periodically conduct tests in order to assess the safety of the existing nuclear power plants by applying impact type loads to the structures. In this study, a sample third-generation nuclear reactor building has been selected to assess its seismic behavior and to observe the crack propagations of the prestressed outer containment. First, a 3D model has been set up using ABAQUS finite element program. Afterwards, modal analysis is conducted to determine the mode shapes. Nonlinear dynamic time history analyses are then followed using an artificial strong ground motion which is compatible with the mean design spectrum of the previously selected ground motions that are scaled to Eurocode 8 Soil type B design spectrum. Results of the conducted nonlinear dynamic analyses are considered in terms of stress distributions and crack propagations.

Keywords: nuclear reactor building, prestressed containment, nonlinear dynamic time history analysis, ABAQUS 3D finite element model

1. Introduction

Large growth of energy consumption is an essential issue that shall be considered, especially for developing countries. Nuclear power industry seems to take a key role over the continuously increasing energy demand. First known nuclear power station for the generations of electricity was built for commercial use in Russia in 1954, with an output of 5 MW(e). Afterwards, USA constructed a pressurized light water reactor (PWR) with 60 MW(e) which first produced electricity. Many countries over the world followed this first attempt and a rapid growth of nuclear energy production started. Though renewable energy seems to take role in the approaching era, it is estimated that the percentage of the generated electricity from these sources will not be deemed to nuclear power plants. It can be forecasted that up to 2050, nuclear energy and nuclear power plants (NPP) will play a major role in producing electricity worldwide.

Nowadays, supporters of nuclear power admit that this is indeed a technology which is more expensive than its counterparts. Throughout its development, starting from the 1980s, the sector improved the design of reactors. Furthermore, the scientists working in this era aimed to overcome human error or equipment malfunctions. These resulted more robust, fuel efficient and advanced reactor designs where the cost of the construction is also reduced.

Next generation nuclear reactors may be classified into two broad categories: evolutionary and revolutionary [1]. While the Generation III and III+ types belong to the former, Generation IV reactors fell into the latter category. Crimello [2] reported that the design of Generation III type nuclear reactors based on minimizing the risk and thus increase safety. Advanced Boiling and Pressurized Water reactors and Enhanced CANDU 6 are all examples to this type. Passive safety features are incorporated into Generation III type reactors to reach the revolutionary reactors Generation III+. Some examples to these reactors may be: AP1000, VVER 1200 and APR 1400.

Generally, an NPP is composed of five principal buildings: the nuclear island, the annex building, the turbine building, the diesel generator building and radwaste buildings. The nuclear island consists of the containment building, the shield building and the auxiliary building. The containment is one of the most important components of an NPP because it serves as the final barrier under postulated accident conditions.

In the last decade, there is a keen interest on modeling the reactor buildings by using simplified 2D lumped-mass stick model [3–7]. The results of the analysis may not be considered as accurate as 3D finite element modeling, although the latter one is more time consuming. In the 3D approach, a containment building is generally modeled using either by shell or 3D brick elements [8–10]. In a recent approach to save time, 3D lumped mass stick models were developed and used in the literature to represent the seismic behavior of containment building [11]. This approach is mostly preferred when the coupling effects between the containment building and auxiliary buildings are considered. Improvements in computational efficiency increased the use of 3D finite element models which are capable of defining high levels of structural detail.

While a possible damage in an infra-structure can be repaired or retrofitted within mean time, a possible damage in an NPP may cause catastrophic damage. The massive damage in an NPP may be observed when an internal accident happens such as loss of coolant accident (LOCA) or when an external event (airplane crash, earthquake, explosions etc.) happens. To overcome these deficiencies the majority of the latest NPP's are constructed with double containment. Further, in advance, the containment structure constitutes an ultimate barrier against the dissemination of fissile products towards the general public [12].

Kwak and Kim [13] highlighted that most of the recent containments are composed of a dome, a wall and a foundation which are laterally prestressed. The International Federation for Structural Concrete (fib) [14] reported that the response of these shell-type concrete structures due to external events should be experimentally and analytically studied to evaluate their safety. There are numerous experiments in the literature which can be regarded as representative experimental studies. Sandia National Laboratories [15] brought foreword the cost and time-consuming features of the conducted experiments and emphasized that these costly experiments often do not precisely simulate the loading and support conditions of the actual structure.

In this study, seismic behavior of the prestressed outer containment of a third-generation nuclear reactor building is evaluated. Since Turkish Building Earthquake Code (TBEC) [16] does not cover the earthquake resistant design of nuclear structures, international standards such as ASCE 4-16 [17] and ASCE 43-05 [18] have been adopted for the analysis. First, a 3D model of the outer containment vessel has been set up using ABAQUS finite element program. Afterwards, modal analysis is conducted to determine the mode shapes and followed by nonlinear dynamic time history analysis using previously selected ground motions to determine the stress distribution and crack propagation.

2. Details of the 3D analytical model of case-studied nuclear reactor building

Both the Sandia National Laboratories and United States Nuclear Regulatory Commission (USNRC) conducted test programs concerning about the seismic performance of prestressed containments under severe accident conditions. The experiment model was 1:4 prestressed containment and it was conducted in two phases. Determination of the ultimate load capacity and seismic response were carried in the first phase, whereas beyond design basis response was the aim of second phase. More details about this experiment may be found in Hessheimer et al. [19, 20].

A sectional view of the considered test building is illustrated in **Figure 1**. The structure has an overall height of 16.4 m, the inner radius of the dome is 5.37 m and the cylinder wall has a thickness of 0.325 m. The bottom part of the structure is fixed, gallery and buttresses are incorporated into the system. Prestressing force is applied by using horizontal and vertical tendons which are anchored at the buttresses and gallery, respectively. Around the gallery and buttresses steel rebars are used to enhance the capacity of the double-shell containment building.

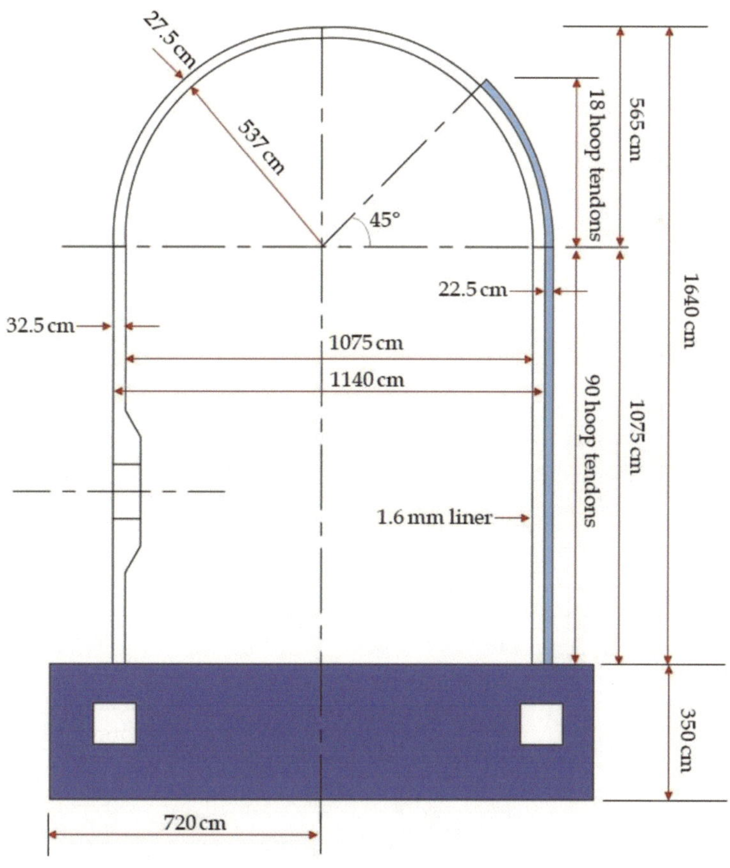

Figure 1.
Sectional view of the considered test building.

Figure 2.
Illustration of the three-dimensional model of the reactor building.

The three-dimensional model of the test structure has been modeled using shell elements. Smeared steel layers have been used to simulate both the steel reinforcements and prestressing tendons. The created three-dimensional model has been illustrated in **Figure 2**. To satisfy the experiment conditions the containment is assumed as fix, thus boundary conditions are assigned to the structure at the bottom. In literature two alternatives are valid to define the prestressing force: assigning an initial stress value and altering with the temperature of the structure. Here the former methodology is applied to properly simulate the prestressing force. The abrupt changes around the openings are neglected. Incremental load steps are used to define the internal pressure values.

Yielding of reinforcing steel and prestressed tendons has been selected as the criteria of failure in any location in the containment. The total mass of the structure has been calculated as 2,956,294 kg and the results of the eigenvalue analysis are summarized in **Table 1**.

Mode No	Eigenvalue	Frequency	
		(rad/time)	(cycles/time)
1	0.36315	0.60262	9.5910E-02
2	0.52506	0.72461	0.11533
3	4.6031	2.1455	0.34146
4	6.6812	2.5848	0.41138
5	12.831	3.5820	0.57009
6	18.840	4.3405	0.69081
7	31.426	5.6059	0.89220
8	40.622	6.3735	1.0144
9	43.410	6.5887	1.0486
10	45.345	6.7339	1.0717

Table 1.
Eigenvalue output.

3. Constitutive material models

3.1 Constitutive model for concrete

The constitutive relation of concrete material is simulated by concrete damaged plasticity model (CDP) of ABAQUS, which is proposed by Lubliner et al. [21] and by Lee and Fenves [22]. This model is preferred by the scientists owing the fact that the model is capable of defining the concrete behavior under cyclic or dynamic loading properly. Further, the material model aims to properly simulate the effects of residual damage. Both the tensile and compressive damages are taken into account in the CDP model. Main contents of the CDP model have been discussed and summarized in the followings.

a. *Decomposition of the strain rate.*

This is done by assuming an additional strain rate decomposition for the rate independent model which is defined by Eq. (1).

$$\acute{\varepsilon} = \acute{\varepsilon}^{el} + \acute{\varepsilon}^{pl} \tag{1}$$

where $\acute{\varepsilon}$ is the total strain rate, $\acute{\varepsilon}^{el}$ is the elastic part of the strain rate, and $\acute{\varepsilon}^{pl}$ is the plastic part of the strain rate.

b. *Stress–strain relations.*

This is dominated by scalar damaged elasticity value which can easily be calculated using Eq. (2).

$$\sigma = (1-d)D_0^{el} : \left(\acute{\varepsilon} - \acute{\varepsilon}^{pl}\right) = D^{el} : \left(\acute{\varepsilon} - \acute{\varepsilon}^{pl}\right) \tag{2}$$

where D_0^{el}; $D^{el} = (1-d)D_0^{el}$; and d, show the initial stiffness, the degraded elastic stiffness and the scalar stiffness degradation variable, respectively. The scalar stiffness degradation may be associated with the failure mechanism of the concrete which has a range $0 \le d \le 1$. Thus, the effective stress is calculated using Eq. (3), and the Cauchy stress by Eq. (4).

$$\overline{\sigma}^{def} = D_0^{el} : \left(\dot{\varepsilon} - \dot{\varepsilon}^{pl}\right) \tag{3}$$

$$\sigma = (1-d)\acute{\sigma} \tag{4}$$

$(1-d)$ in Eq. (4) is the factor that represents the effective load carrying area. When $d = 0$, reduction in the elastic stiffness is not expected and it is the condition where $\acute{\sigma} = \sigma$. When $d \neq 0$, damage is possible and use of the effective stress value will give more robust results. Hence, in problems dealing with plasticity $\acute{\sigma}$ value is preferred.

c. *Hardening variables.*

It is also possible to relate d value by a set of hardening variables, $\tilde{\varepsilon}^{pl}$, and $\overline{\sigma}$; such, $d = d\left(\overline{\sigma}, \tilde{\varepsilon}^{pl}\right)$.$\tilde{\varepsilon}_t^{pl}$ and $\tilde{\varepsilon}_c^{pl}$ are referred to define damage states in tension and compression, respectively. The governing equation of the hardening variables is given in Eq. (5).

$$\tilde{\varepsilon}^{pl} = \begin{bmatrix} \tilde{\varepsilon}_t^{pl} \\ \tilde{\varepsilon}_c^{pl} \end{bmatrix}; \dot{\tilde{\varepsilon}}^{pl} = h(\overline{\sigma}, \tilde{\varepsilon}^{pl}).\dot{\varepsilon}^{pl} \tag{5}$$

These variables control the yield surface, the degradation of the elastic stiffness and the dissipated fracture energy.

d. Yield function.

The yield function, $F(\overline{\sigma}, \tilde{\varepsilon}^{pl})$, represents a surface in effective stress space, which determines the states of failure or damage. For the inviscid plastic-damage model, $F(\overline{\sigma}, \tilde{\varepsilon}^{pl}) \leq 0.F$ can also be defined in terms of $\overline{\sigma}$, as in Eq. (6).

$$F(\overline{\sigma}, \tilde{\varepsilon}^{pl}) = \frac{1}{1-\alpha}\left(\overline{q} - 3\alpha\overline{p} + \beta(\tilde{\varepsilon}^{pl})\langle\hat{\overline{\sigma}}_{max}\rangle - \gamma\langle-\hat{\overline{\sigma}}_{max}\rangle\right) - \overline{\sigma}_c(\tilde{\varepsilon}_c^{pl}) \leq 0 \tag{6}$$

α and γ in Eq. (6) are dimensionless constants. $\overline{p}$ and $\overline{q}$ are the effective hydrostatic pressure and the Von Mises equivalent effective stress values, respectively. Further these parameters can be calculated by Eqs. (7) and (8). The deviatoric stress may then be expressed using Eq. (9). Here $\hat{\overline{\sigma}}_{max}$ refers to the maximum eigenvalue of the effective stress value.

$$\overline{p} = -\frac{1}{3}\overline{\sigma} : I \tag{7}$$

$$\overline{q} = \sqrt{\frac{3}{2}\overline{S} : \overline{S}} \tag{8}$$

$$\overline{S} = \overline{p}I + \overline{\sigma} \tag{9}$$

When $\hat{\overline{\sigma}}_{max} = 0$ is the case, Eq. (6) reduces to the Drucker-Prager yield statement. Then the α coefficient defined in (6) may be determined using the initial equibiaxial (σ_{b0}) and uniaxial compressive yield stress (σ_{c0}) as presented in Eq. (10). Experimental results revealed the fact that $0.08 \leq \alpha \leq 0.12$ [21].

$$\alpha = \frac{\sigma_{b0} - \sigma_{c0}}{2\sigma_{b0} - \sigma_{c0}} \tag{10}$$

The $\beta(\tilde{\varepsilon}^{pl})$ is then defined using these α parameters as given in Eq. (11). Here $\overline{\sigma}_t$ and $\overline{\sigma}_c$ are the effective tensile and compressive cohesion stresses, respectively.

$$\beta(\tilde{\varepsilon}^{pl}) = \frac{\overline{\sigma}_c(\tilde{\varepsilon}_c^{pl})}{\overline{\sigma}_t(\tilde{\varepsilon}_t^{pl})}(1-\alpha) - (1+\alpha) \tag{11}$$

e. Flow rule.

Defining a flow potential value G, plastic flow rule may be expressed using Eq. (12). Here $\dot{\lambda}$ is a plastic multiplier which always has a positive value. Following the Kuhn-Tucker relations, which are $\dot{\lambda}F = 0; \dot{\lambda} \geq 0; F \leq 0$, the G parameter is defined in the $\overline{\sigma}$ space.

$$\dot{\varepsilon}^{pl} = \dot{\lambda}\frac{\partial G(\overline{\sigma})}{\partial \overline{\sigma}} \tag{12}$$

Using the Drucker-Prager hyperbolic function, the G parameter may be defined using Eq. (13). Here ψ is the dilation angle at high confining pressure; σ_{t0} is the uniaxial tensile stress at failure; and ϵ refers to an eccentricity parameter.

$$G = \sqrt{(\epsilon\sigma_{t0}tan\psi)^2 + \overline{q}^2} - \overline{p}tan\psi \tag{13}$$

f. Damage and stiffness degradation.

This section will be considered for both uniaxial and cyclic loading protocols.

- Uniaxial loading

The assumption relies on the fact that the uniaxial stress–strain curves can be represented in terms of stress and plastic strain values using Eq. (14).

$$\sigma_t = \sigma_t\left(\tilde{\varepsilon}_t^{pl}, \dot{\tilde{\varepsilon}}_t^{pl}, \theta, f_i\right), \sigma_c = \sigma_c\left(\tilde{\varepsilon}_c^{pl}, \dot{\tilde{\varepsilon}}_c^{pl}, \theta, f_i\right) \tag{14}$$

In Eq. (14) the subscripts t and c refer to tension and compression, respectively; $\dot{\tilde{\varepsilon}}_t^{pl}$ and $\dot{\tilde{\varepsilon}}_c^{pl}$ are the equivalent plastic strain rates, $\tilde{\varepsilon}_t^{pl} = \int_0^t \dot{\tilde{\varepsilon}}_t^{pl} dt$ and $\tilde{\varepsilon}_c^{pl} = \int_0^t \dot{\tilde{\varepsilon}}_c^{pl} dt$ are the equivalent plastic strains, θ is the temperature, and f_i, $(i = 1, 2, ..)$ are other predefined field variables. Under uniaxial loading conditions the effective plastic strain rates are given as: $\dot{\tilde{\varepsilon}}_t^{pl} = \dot{\varepsilon}_{11}^{pl}$, in uniaxial tension and $\dot{\tilde{\varepsilon}}_c^{pl} = -\dot{\varepsilon}_{11}^{pl}$, in uniaxial compression (ABAQUS 2020). The response of concrete to uniaxial loading is as illustrated in **Figure 3**. The left illustration is for when the system is under effect of compression only and the right one corresponds to tension case.
To properly evaluate the degraded response of concrete two independent damage variables which are assumed as functions of the strain values are introduced: d_t and d_c. These variables can be calculated following Eqs. (15) and (16). If there is no damage in the material then these variables are assumed to be equal to zero, on the contrary these variables are taken as one when the material is fully damaged.

$$d_t = d_t\left(\tilde{\varepsilon}_t^{pl}, \theta, f_i\right), (0 \le d_t \le 1) \tag{15}$$

$$d_c = d_c\left(\tilde{\varepsilon}_c^{pl}, \theta, f_i\right), (0 \le d_c \le 1) \tag{16}$$

Defining E_O as the initial modulus of the material, the stress values may be determined using Eq. (17).

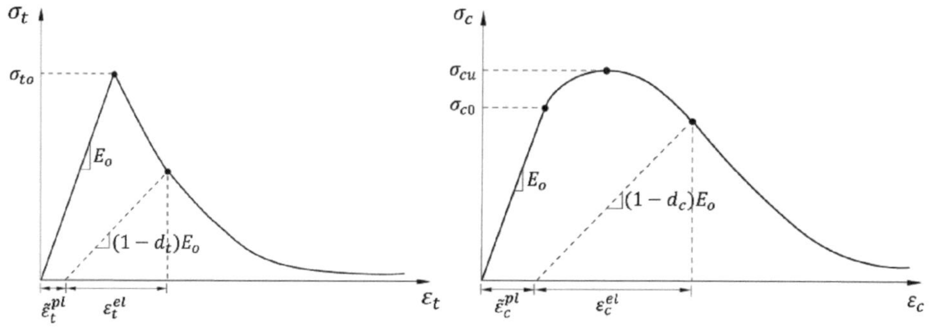

Figure 3.
Response of concrete to uniaxial loading.

$$\sigma_t = (1-d)E_O\left(\varepsilon_t - \tilde{\varepsilon}_t^{pl}\right), \sigma_c = (1-d)E_O\left(\varepsilon_c - \tilde{\varepsilon}_c^{pl}\right) \tag{17}$$

When tension loading protocol is the case, cracks propagate in a direction transverse to the stresses. Further, cracks reduce the load bearing capacity of the material and this increases the $\overline{\sigma}$ value. On the contrary, when compressive loading is the case, cracks propagate in a direction parallel to the stresses. This minimize the effect of cracks. Naming $\overline{\sigma}_t$ and $\overline{\sigma}_c$ as the effective uniaxial cohesion stresses values at tension and compression, respectively, Eqs. (18) and (19) may be applicable to determine the size of the yield surface.

$$\overline{\sigma}_t = \frac{\sigma_t}{(1-d_t)} = E_0\left(\varepsilon_t - \tilde{\varepsilon}_t^{pl}\right), \tag{18}$$

$$\overline{\sigma}_c = \frac{\sigma_c}{(1-d_c)} = E_0\left(\varepsilon_c - \tilde{\varepsilon}_c^{pl}\right) \tag{19}$$

- Cyclic loading

The interaction of the cracks makes this type of loading protocol more complex to understand the behavior. Previous experiments revealed the fact that there is indeed some recovery of the stiffness and this is named as "the stiffness recovery effect (SRE)". This is a significant behavior of concrete under cyclic loading, especially when the load changes from tension to compression. To correlate *d* and *E* values, Eq. (20) has been proposed for the CDP model.

$$E = (1-d)E_0 \tag{20}$$

It is previously shown that d may be related with d_t and $d_c$values. When cyclic loading protocol is applied ABAQUS follows the relation given in Eq. (21). In Eq. (21) s_t and s_c are stress functions which includes the SRE associated with stress reversals.

$$(1-d) = (1-s_t d_c)(1-s_c d_t), 0 \le s_t, s_c \ge 1 \tag{21}$$

Introducing ω_t and ω_c material weight factors responsible from the tensile and compressive stiffness recovery, Eqs. (22) and (23) is proposed. The constant r in Eqs. (22) and (23) is determined from Eq. (24).

$$s_t = 1 - \omega_t r * (\overline{\sigma}_{11}); 0 \le \omega_t \le 1 \tag{22}$$

$$s_c = 1 - \omega_c (1 - r * (\overline{\sigma}_{11})); 0 \le \omega_c \le 1 \tag{23}$$

$$r * (\overline{\sigma}_{11}) = H(\overline{\sigma}_{11}) = \begin{cases} 1 \; if \; \overline{\sigma}_{11} > 0 \\ 0 \; if \; \overline{\sigma}_{11} < 0 \end{cases} \tag{24}$$

This observation in concrete may be accepted as a proof of SRE when cracks close as the load goes from tension to compression. When vice versa is valid, then the tensile stiffness is not recovered. This is ensured in ABAQUS by assuming $\omega_t = 0$ and $\omega_c = 1$. Cyclic load behavior is illustrated in **Figure 4**.

3.2 Constitutive model for steel

Isotropic hardening model is used to simulate the elastic–plastic steel behavior. As well as the rebars, prestressed tendons are defined using steel material. This indicates that the yield surface changes in all directions. To define the yield function $f(\sigma)$

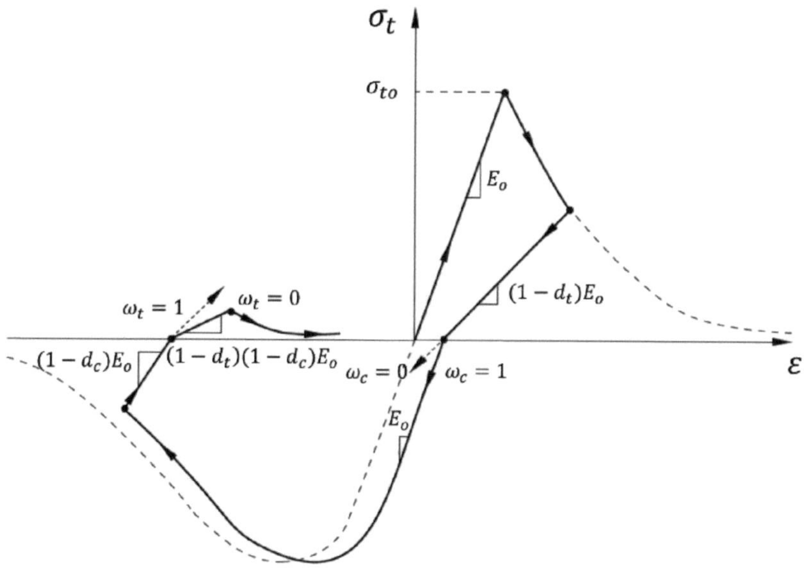

Figure 4.
Cyclic loading protocol for $\omega_t = 0$ and $\omega_c = 1$.

properly, equivalent stress σ^0 is introduced first. This stress value is indeed a function of strain and temperature, as presented in Eq. (25). Here θ is the considered temperature and $\tilde{\varepsilon}^{pl}$ gives the value of the equivalent plastic strain, given in Eq. (26).

$$f(\sigma) = \sigma^0\left(\tilde{\varepsilon}^{pl}, \theta\right) \tag{25}$$

$$\sigma^0 \dot{\tilde{\varepsilon}}^{pl} = \sigma : \dot{\varepsilon}^{pl} \tag{26}$$

Rice [23] brought forward the use of this model especially when the Bauschinger effect is forceful. The strain rate decomposition may be calculated from Eq. (27). When Eq. (27) is integrated through the contour of the yield surface Eq. (28) is obtained.

$$d\varepsilon = d\varepsilon^{el} + d\varepsilon^{pl} \tag{27}$$

Parameters	Values
Elastic modulus (MPa)	3.6×10^4
Ultimate tensile strength (MPa)	2.85
Ultimate compressive strength (MPa)	38.5
Poisson ratio	0.2
Density (kg/m^3)	2400
Thermal expansion coefficient	1×10^{-5}
Viscosity parameter	0.005
Dilation angle	36.31
Eccentricity	0.1
σ_{b0}/σ_{co}	1.16
K_C	0.667

Table 2.
Parameters of the concrete material.

Parameters	Rebars	Tendons
Elastic modulus (MPa)	2×10^5	1.95×10^5
Yield strength (MPa)	486.55	1860
Yield strain (m/m)	0.002613	0.008745
Poisson ratio	0.3	0.3
Density (kg/m^3)	7850	7850
Thermal expansion coefficient	1×10^{-5}	1×10^{-5}

Table 3.
Parameters of the steel material.

$$\varepsilon = \varepsilon^{el} + \varepsilon^{pl} \tag{28}$$

The elasticity can be then written in terms of two temperature-dependent material parameters. These parameters are generally selected as the bulk modulus, K, and the shear modulus, G, which are given Eqs. (29) and (30).

$$K = \frac{E}{3(1 - 2\upsilon)} \tag{29}$$

$$G = \frac{E}{2(1 + \upsilon)} \tag{30}$$

The constitutive material parameters used in this study for concrete and steel are as given in **Tables** 2 and **3**, respectively.

4. Seismic performance assessment of the selected reactor containment building

Two basic regulations for NPP's may be named as the 10 CFR 50 and 10 CFR 100 which are published by the Nuclear Regulatory Commission (USNRC) [24, 25]. The former one covers the licensing issues and the latter one explains the steps of evaluating a license. In the latter regulation, two basic earthquakes are defined when performing a seismic hazard assessment: safe shutdown earthquake (SSE) and operating basis earthquake (OBE). While SSE is considered to be the maximum earthquake which could occur at the investigation site, OBE is defined as the earthquake which could be expected to occur at the site during the lifetime of the plant. Further in the regulation, it is stated that the maximum vibratory ground acceleration of the OBE must be at least 33% of the maximum vibratory ground acceleration of the SSE [24, 25].

Nowadays, the methodology for risk analysis involves the use of component fragility curves developed using ground-motion parameters. To obtain the fragility curves the capacity and the demand parameters should be evaluated at first instance. It is a well-known fact that failure of both structural and nonstructural components of an NPP are much more involved with the structural response than the ground parameters.

The FEMA P-58-1 Guidelines [26] provide a basis to improve the risk assessment procedure for NPPs. The guideline develops next-generation tools and new approaches for performance assessment of buildings, with a focus on measuring performance in terms of direct economic loss, casualties and downtime. The FEMA P-58 Guidelines [26] present procedures for performance assessment using a probabilistic framework, which provides a robust methodology to integrate hazard curves, component fragility curves and consequence functions and to capture the dispersions in each of these elements for evaluating the performance of a building. Importantly, the fragility curves used in the analysis are defined in terms of structural response parameters.

SDC	Limit state			
	A Large permanent distortion (short of collapse)	B Moderate permanent distortion	C Limited permanent distortion	D Essentially elastic
1	SDB-1A ASCE7	SDB-1B ASCE7	SDB-1C ASCE7	SDB-1D NA
2	SDB-2A ASCE7	SDB-2B ASCE7	SDB-2C NA	SDB-2D NA
3	SDB-3A ASCE 43-05	SDB-3B ASCE 43-05	SDB-3C ASCE 43-05	SDB-3D ASCE 43-05
4	SDB-4A ASCE 43-05	SDB-4B ASCE 43-05	SDB-4C ASCE 43-05	SDB-4D ASCE 43-05
5	SDB-5A ASCE 43-05	SDB-5B ASCE 43-05	SDB-5C ASCE 43-05	SDB-5D ASCE 43-05

Table 4.
Graded approach defined in ASCE 43-05 [18].

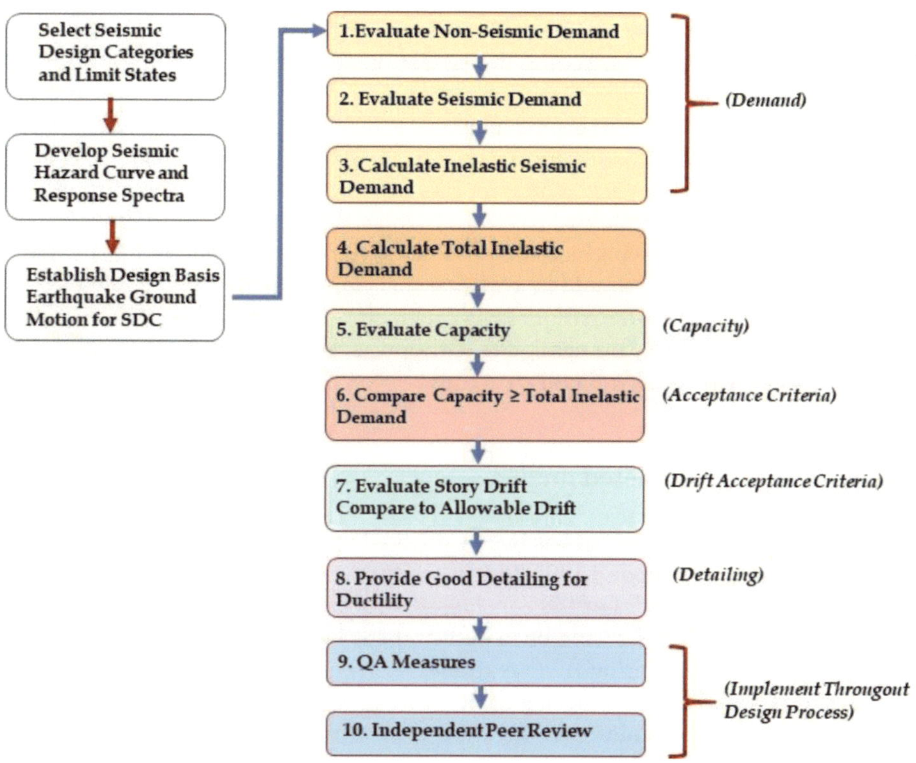

Figure 5.
Seismic design procedure defined in ASCE 43-05 [18] for SDC 3, 4, and 5.

Limit state	Structural deformation limit
A	Large permanent distortion, short of collapse *Significant damage*
B	Moderate permanent distortion *Generally repairable damage*
C	Limited permanent distortion *Minimal damage*
D	Essentially elastic behavior *No damage*

Table 5.
Structural deformation limits for limit state defined in ASCE 43-05 [18].

ASCE 43-05 [18] provides seismic design criteria for structures, systems, and components (SSC) that are used in nuclear facilities. This is executed by using a graded approach where the design criteria are proportional with the relative importance to safety, magnitude of the seismic event and other factors. To achieve this 20 Seismic Design Bases (SDB) are defined, where for each Seismic Design Category (SDC) probabilistic target performance goals are used. This graded approach is summarized in **Table 4**. Further in the same guide for each Limit State (LS) an acceptable level of structural response goal is defined. While International Building Code (IBC) [27] may be followed for SDBs defined by SDC1 and 2; for SDBs defined by SDC 3, 4, and 5, ASCE standard should be followed as illustrated in **Figure 5**. Further in the mentioned code, the Design Basis Earthquake (DBE) is defined by Uniform Hazard Response Spectra (UHRS).

For DBE, Limit State A and D are defined as the intensity of the high and low structural damage, respectively. In other words, Limit State D is where complete elastic behavior is dominant. Damage regions between A and D are called the intermediate levels. The deformation limits associated with each Limit State are described in **Table 5**.

To properly determine the seismic demand, ASCE 43-05 [18] allows use of *linear equivalent static analysis*, *linear dynamic analysis*, *complex frequency response methods*, or *nonlinear analysis.* Moreover, when the fundamental mode is dominant then a single step pushover methodology is allowed while conducting nonlinear analysis. These procedures shall be conducted by following the criteria given in FEMA-356 [28] or ATC-40 [29]. This should be clarified that nonlinear analysis is only permitted when beyond design earthquake is considered for low-rise regular NPPs, where higher mode effects are negligible. Otherwise, elastic models are preferred in the seismic design of safety-related structures.

5. Results of the conducted analysis

The 3D finite element model has been subjected to gravity, wind and snow loads in accordance with Eurocode 1 [30], Section 4. First, to determine the fundamental periods and vibration modes of the considered structure, eigenvalue analyses are performed and eigenvectors have been determined. **Figure 6** shows the results of the modal analysis considering first six modes.

As it can be inferred from **Figure 6**, first three modal results reveal the fact that the dome of the reactor building is the most vulnerable section. The displacement value of the dome is calculated approximately 1 mm for the fundamental mode. When considered the higher modes, stress concentrations around operational gaps are observed.

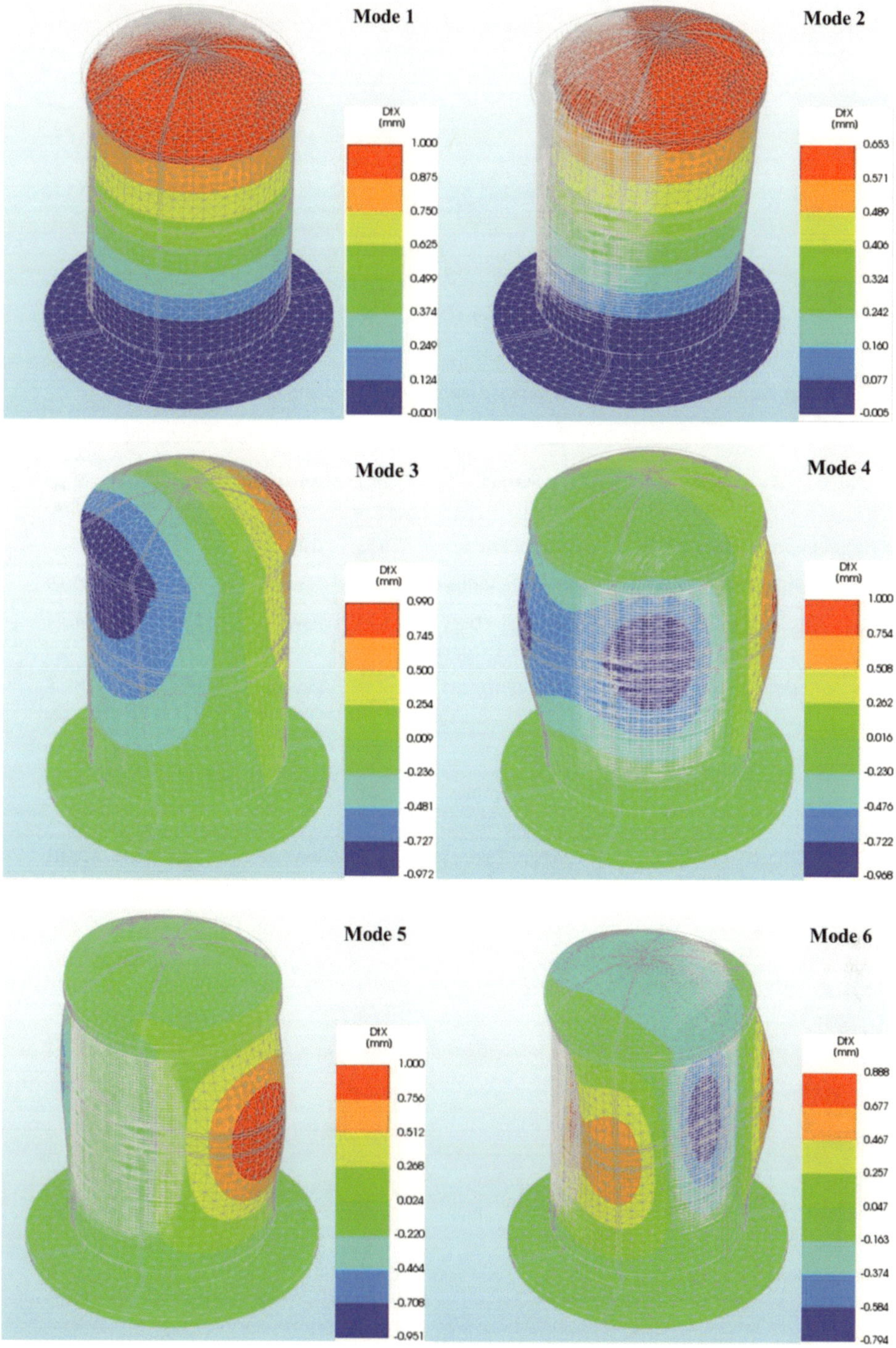

Figure 6.
Calculated mode shapes and periods.

After determining the mode shapes of the outer containment vessel, nonlinear dynamic time history analyses are conducted to assess the seismic behavior of the structure and the stress distribution. For this purpose, PEER database has been used and seven earthquake excitations are first selected in accordance with ATC 58 [31] and scaled to fit Eurocode 8 [32] Soil type B design spectrum. **Figure 7** compares

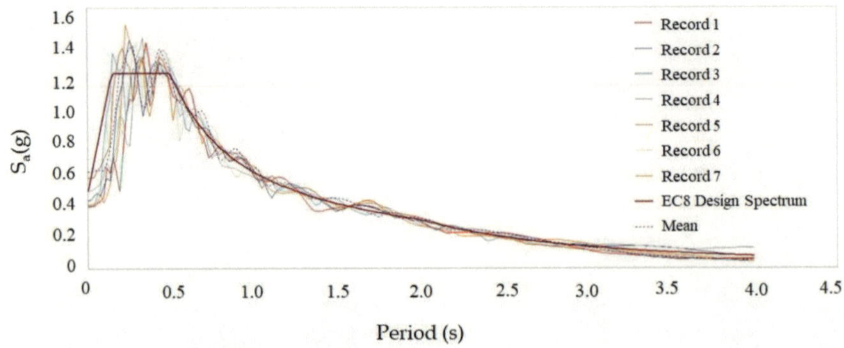

Figure 7.
Comparison of Eurocode 8 design spectrum with the selected ground motions spectra.

No	Earthquake	Station	M_w	Mechanism	V_{s30} (m/s)	R_{jb} (km)	R_{rup} (km)
1	Imperial Valley (1979)	Cerro Prieto	6.53	Strike slip	471.53	15.19	—
2	Irpinia Italy (1980)	Rionero In Vulture	6.2	Normal	574.88	22.68	22.69
3	Loma Prieta (1989)	Coyote Lake Dam - Southwest Abutment	6.93	Reverse Oblique	561.43	19.97	20.34
4	Chi-Chi Taiwan (1999)	CHY010	7.63	Reverse Oblique	538.69	19.93	19.96
5	Chi-Chi Taiwan (1999)	TCU075	6.3	Reverse	573.02	24.34	26.31
6	Landers (1992)	North Palm Springs Fire Sta #36	7.28	Strike slip	367.84	26.95	26.95
7	Chuetsu-oki Japan (2007)	Matsushiro Tokamachi	6.8	Reverse	640.14	18.16	25.03

Table 6.
Details of the selected strong ground motions.

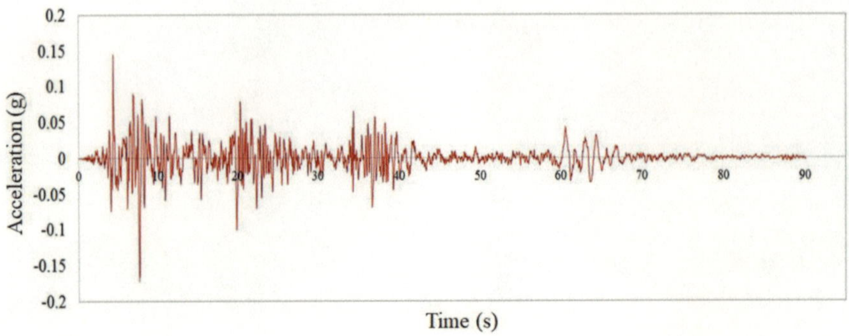

Figure 8.
Artificial ground motion.

Eurocode 8 [32] design spectrum with the spectra of the selected ground motions. The details of the selected strong ground motions are presented in **Table 6**. Following, an artificial time history record which is compatible with the mean spectrum has been generated and used in the nonlinear dynamic analyses. **Figure 8** represents the artificial ground motion.

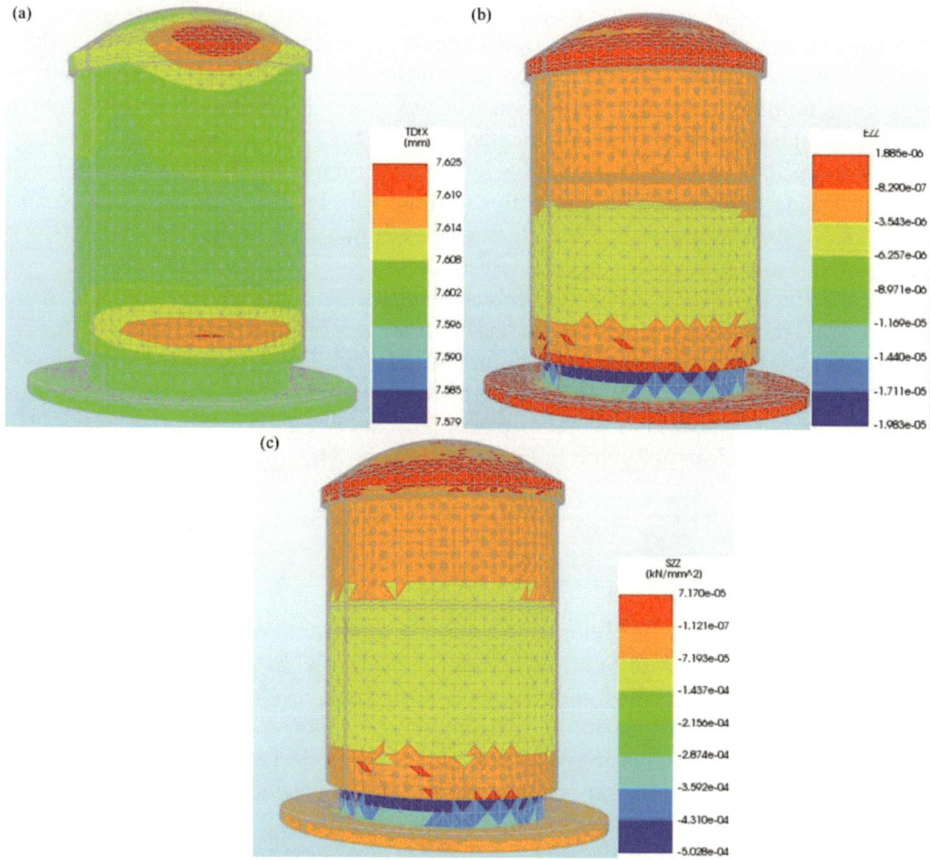

Figure 9.
Nonlinear dynamic time history results of the outer containment: (a) total displacement, (b) total deformation and (c) total stress concentration.

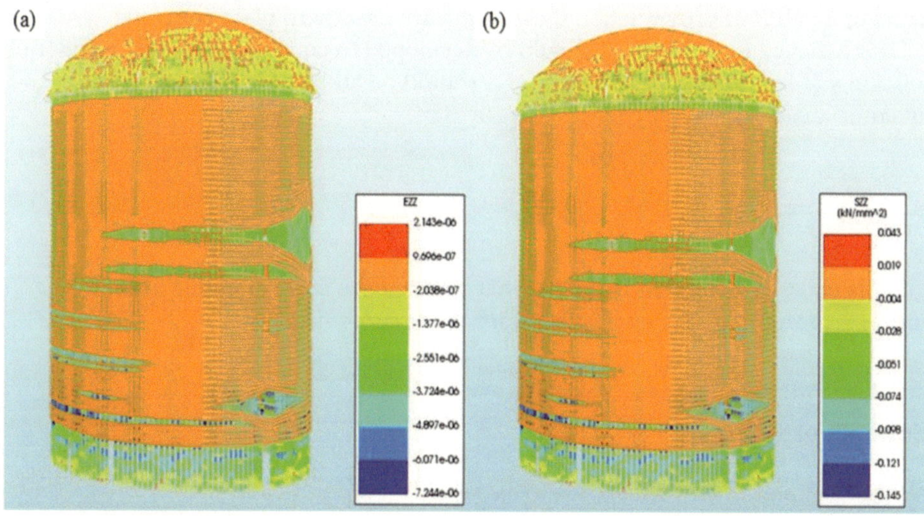

Figure 10.
Nonlinear dynamic time history results of the prestressed tendons: (a) total displacement and (b) total stress concentration.

Modified Newton–Raphson iterative procedure has been followed during the dynamic analysis. ABAQUS offers several convergence norms. Besides stopping the iteration in case of convergence, the iteration process is also stopped if a specified maximum number of iterations has been reached or if the iteration obviously leads to divergence. The detection of divergence is based on the same norms as the detection of convergence. A preferred way to check the convergence is the energy norm.

As the consequence of dynamic analysis total displacement, total deformation and total stress concentration values are illustrated in **Figure 9**. It should also be highlighted that results of the modal analysis are in good correlation with the dynamic results and the most vulnerable section is assessed as the dome of the structure. In **Figure 9(a)**, total displacement graph has been presented. The maximum displacement value has been calculated as 7.625 mm at the roof. The deformation graph, which is represented in **Figure 9(b)**, is compatible with the stress concentration that is represented in **Figure 9(c)**. Total displacement and stress concentration graphs of the prestressed tendons are given in **Figure 10(a)** and **(b)**, respectively.

6. Conclusions

In this study, seismic behavior of the outer containment of a sample reactor building has been analyzed. First, details of the 3D model have been presented, and material models are described in detail. Following that, results of the eigenvalue analysis are discussed. Then, nonlinear time history analyses are conducted using the artificial record which is compatible with the mean design spectrum of the selected ground motions that are previously scaled to Eurocode 8 Soil Type B design spectrum.

The primary outcome of the study is that the most critical section of a reactor building may be assessed as the dome of the structure. More care should be given while designing the dome. In the recent literature, to overcome the deficiencies of prestressed concrete, use of fiber reinforced concrete is proposed due to the fact that these fibers in plain concrete directly effects both the ultimate capacity and post-cracking behavior of a conventional prestressed containment. It may also be interpreted from these ongoing studies that the structural performance of components of an NPP improves when these fibers are used with plain concrete.

It is believed that the study should be developed to consider inelastic behavior of soil and soil-structure interaction effects should be taken into account in the dynamic analysis.

Acknowledgements

This paper is generated from the PhD study of the corresponding author. The author is grateful for the endless support of her supervisor.

Conflict of interest

The authors declare no conflict of interest.

Author details

Evrim Oyguc*, Abdul Hayır and Resat Oyguc
Istanbul Technical University, Istanbul, Turkey

*Address all correspondence to: eoyguc@itu.edu.tr

References

[1] Kessides IN. The future of the nuclear industry reconsidered: Risks, uncertainties, and continued promise. Energy Policy. 2012;**48**:185-208

[2] Crimello R. Achievements and prospects for advanced reactor design and fuel cycles. IAEA – Scientific forum. 2004. In: Nuclear Fuel Cycles Issues and Challenges. 2004

[3] Varma V, Reddy GR, Vaze KK, Kushwaha HS. Simplified approach for seismic analysis of structures. International Journal of Structural Stability and Dynamics. 2002;**2**(2): 207-225

[4] Huang YN, Whittaker AS, Luco N. Seismic performance assessment of base-isolated safety-related nuclear structures. Earthquake Engineering and Structural Dynamics. 2010;**39**(13): 1421-1442

[5] Hirama T, Goto M, Hasegawa T, Kanechika M, Kei T, Mieda T, et al. Seismic proof test of a reinforced concrete containment vessel (RCCV): Part 1: Test model and pressure test. Nuclear Engineering and Design. 2005; **235**(13):1335-1348

[6] Leonardo TS, Richard SO, Sener T, Diego PR. Finite element modeling of the AP1000 nuclear island for seismic analyses at generic soil and rock sites. Nuclear Engineering and Design. 2007; **237**(12-13):1474-1485

[7] Jeremic B, Tafazzoli N, Ancheta T, Orboviæ N, Blahoianu A. Seismic behavior of NPP structures subjected to realistic 3D, inclined seismic motions, in variable layered soil/rock, on surface or embedded foundations. Nuclear Engineering and Design. 2013;**265**:85-94

[8] Bily P, Kohoutkova A. An estimation of the effect of steel liner on the ultimate bearing capacity of prestressed concrete containment. Nuclear Engineering and Design. 2018;**328**:197-208

[9] Park J, Park N, Lee S, Park Y, Choi Y. Seismic analysis of the APR1400 nuclear reactor system using a verified beam element model. Nuclear Engineering and Design. 2017;**313**:108-117

[10] Ambrosini D, Codina RH, Curadelli O, Martinez CA. Structural analysis of the CAREM-25 nuclear power plant subjected to the design basis accident and seismic loads. Annals of Nuclear Energy. 2017;**108**:42-56

[11] Tunon-Sanjur L, Orr RS, Tinic S, Ruiz DP. Finite element modeling of the AP1000 nuclear island for seismic analyses at generic soil and rock sites. Nuclear Engineering and Design. 2007; **237**:1474-1485

[12] De Boeck B. A review of containment accidents. Nuclear Engineering Design. 1993;**145**:279-288

[13] Kwak HG, Kim JH. Numerical models for prestressing tendons in containment structures. Nuclear Engineering and Design. 2006;**236**:1061-1080

[14] fib Task Group on Containment Structures. Nuclear containments. In: International Federation for Structural Concrete (Fib). 2001

[15] Sandia National Laboratories. Pretest round Robin analysis of a prestressed concrete containment vessel model. U.S. Nuclear Regulatory Commission (NRC) and nuclear power engineering corporation (NUPEC). In: NUREG/CR-6678. 2000

[16] TBEC TBEC. Ministry of Public Work and Settlement. Turkey: Ankara; 2018

[17] ASCE 4-16. Seismic Analysis of Safety-Related Nuclear Structures.

American Society of Civil Engineers. 2017. 1801 Alexander Bell Drive, Virginia, U.S.A.

[18] ASCE 43-05. Seismic Design Criteria for Structures, Systems, and Components in Nuclear Facilities, American Society of Civil Engineers. 2005. Alexander bell drive, Virginia. In: U.S.A. 1801

[19] Hessheimer MF, Dameron RA. Containment Integrity Research at Sandia National Laboratories – An Overview, Technical Report NUREG/CR-6906; SAND2006-2274P. Sandia National Laboratories and USNRC, Washington. In: USA. 2006

[20] Hessheimer MF, Klamerus EW, Lambert LD, Rightley GS, Dameron RA. Over Pressurization Test of a 1:4-Scale Prestressed Concrete Containment Vessel Model, Technical Report National/CR-6810; SAND2003-0840p. Washington, USA: Sandia National Laboratories and USNRC; 2006

[21] Lubliner J, Oliver J, Oller S. Onate E. a plastic-damage model for concrete. International Journal of Solids and Structures. 1989;**25**(3):229-326

[22] Lee J, Fenves GL. Plastic-damage model for cyclic loading of concrete structures. Journal of Engineering Mechanics. 1998;**124**(8):892-900

[23] Rice JR. Continuum Mechanics and Thermodynamics of Plasticity in Relation to Microscale Deformation Mechanisms. Constitutive Equations in Plasticity: Argon, MIT Press, Cambridge, Massachusettes; 1975

[24] USNRC, Individual Plant Examination of External Events, (IPEEE) for Severe Accident Vulnerabilities, Generic Letter No. 88-20, Supplement 4. 1991. US Nuclear Regulatory Commission.

[25] USNRC. Procedural and Submittal Guidance for the Individual Plant Examination of External Events (IPEEE) for Severe Accident Vulnerabilities, NUREG-1407. 1991. US Nuclear Regulatory Commission.

[26] FEMA P-58-1. Seismic Performance Assessment of Buildings Volume 1-Methodology. Applied Technology Council. California: Redwood City; 2012

[27] IBC. International Building Code. International Code Council. Country Club Hills. In: IL. 2003

[28] FEMA 356. Prestandard and Commentary for the Seismic Rehabilitation of Buildings. American Society of Civil Engineers. Virginia: Reston; 2000

[29] ATC 40. Seismic Evaluation and Retrofit of Concrete Buildings Volume 1. Applied Technology Council. California: Redwood City; 1996

[30] Eurocode 1. Design of structures for earthquake resistance. Part 1. General Rules. Specific Rules, Seismic Actions and Rules for Buildings. Belgium: European Committee for Standardization, Bruxel; 2008

[31] ATC 58. Next- Generation Performance-Based Seismic Desigb Guidelines,Program Plan for New and Existing Buildings. FEMA 445. California, US: Redwood City; 2006

[32] Eurocode 8. Design of structures for earthquake resistance. Part 1. General Rules. Specific Rules, Seismic Actions and Rules for Buildings. Belgium: European Committee for Standardization, Bruxel; 2008

Chapter 3

Corrosion Effect on Bond Loss between Steel and Concrete

Charis Apostolopoulos and Konstantinos Koulouris

Abstract

This chapter is devoted to the effects of steel corrosion on bond relationship between steel and concrete. One of the basic assumptions in design of reinforced concrete members is the perfect steel - concrete bond mechanism, so that strain of reinforcing bar is the same as that of the surrounding concrete and these two different materials act as one. However, corrosion of steel reinforcement consists one of the main durability problems in reinforced concrete members, downgrade the bond behavior and therefore their structural integrity. Corrosion degrades the reinforcement itself, reducing the initial cross-section of the steel bar and its mechanical properties. Furthermore, tensile stresses in surrounding concrete caused due to oxides on the corroded reinforcement, lead to the gradual development of tensile field to the surrounding concrete, with spalling of the cover concrete and loss of bond mechanism as a consequence. In this chapter, an overview of damage of reinforced concrete due to steel corrosion is given, focused on the bond mechanism; factors that play key role in the degree of bonding and, also, proposed models of bond strength loss in correlation with the surface concrete cracking due to corrosion are indicated. To conclude, the ongoing research in this area of interest is presented, based on recent scientific studies.

Keywords: steel corrosion, bond loss, surface cracking, bond strength, predictive model, corrosion damage

1. Introduction

Reinforced concrete consists the most widely used construction material of the existing building stock, providing high bearing capacity in conjunction with low production cost. Due to weakness of plain concrete to withstand tensile forces, steel reinforcing bars are introduced into concrete to enhance its overall mechanical performance. The key consideration so as to ensure that steel and concrete cooperate is the bond mechanism between them. However, corrosion of steel reinforcement constitutes a major degradation factor, which leads to premature aging of RC structures. Recent reports have indicated huge economic impacts due to corrosion damage, since a significant part of the annual budget in many countries is spent on maintenance, repair, and rehabilitation of RC structures [1, 2]. For instance, results of a study conducted by NACE [1] refers that the cost of corrosion is globally estimated to be US$2.5 trillion, approximately 3.4 percent of the global GDP.

It is a common knowledge that corrosion process has initially slow rate in nature, since it takes a long period of more than about 10 or 15 years until the aggressive environmental agents diffuse into concrete and reach the steel reinforcement so as to create the electrochemical cell of corrosion. Concrete cover thickness and a thin passive layer on steel surface, due to high alkalinity of concrete (pH at about 12.5) protect the reinforcement, delaying the penetration and diffusion of corrosive factors. Nevertheless, when aggressive environmental factors, as chloride ions, reach a critical concentration rate, accompanied by reduction of pH below 9, steel reinforcement depassivates and corrosion initiates [3]. Several scientific studies have been conducted, investigating the rate of chlorides' diffusion through the porrosive concrete, aimed at the establishment of a critical value (threshold) of chloride content in order to predict the onset of corrosion [4–6]. The majority of researchers model chloride transport in concrete using the Fick's second law of diffusion, neglecting the chloride interaction with the solid phase [7]. However, there are many uncertainties since many factors influence the rate of chloride penetration into concrete, such as porosity and cracks of concrete, temperature, moisture and salinity of corrosive environment. Hence, due to misinterpretations and various results in literature to date, there is no broadly accepted by the scientific community method of estimating and modeling the onset of corrosion by means of the critical chloride content. Due to the abovementioned, modern international regulations on the design of concrete structures, as BSI EN 206–1 [8], based on long term service life of reinforced concrete, proposed minimum values of cover thickness and concrete classes, depending on the environmental exposure conditions, in order to ensure high protection level of reinforcement against corrosion.

As aggressive agents penetrate and act in limited exposed to corrosion areas rather than the entire length of the reinforcing bars, the corrosion effect is mainly characterized by non-uniformity along the steel bars and is detected by pits on their surface. During corrosion process, steel tends to return to its initial ore form resulting in mass loss and its conversion to iron oxides (rust) on steel surface. Consequences of corrosion damage on steel reinforcement are the reduction of the initial cross-section, resulting in increase of applied mechanical stresses and stress concentration due to pit development, as well as the degradation of its mechanical properties. During the last decades, an effort has been made so as to estimate and quantify the corrosion effect on steel reinforcing bars. The non-uniform distribution of corrosion damage on the steel cross section agreed to be one of the primary cause of mechanical properties' degradation, which has been studied by many researchers [9–13]. Sun [12] and Andisheh et al. [13] tested bare reinforcing bars under monotonic loads, depicting the significant material degradation. Extending the research upon steel corrosion in concrete, experimental studies on embedded reinforcing bars [14–16] indicated more severe corrosion damage, accompanied by narrow pits, which leads to further reduction of mechanical response. Nonetheless, experimental results of both bare and embedded corroded steel reinforcement mainly showed ductility drop rather than reduction of effective stress. To this effect, several studies targeted on the relationship between the degree of corrosion of steel reinforcement and the bearing capacity of corresponding reinforced concrete structures [17–19]. Recently, Kashani et al. [20] presented a state of the art review up to date concerning the current knowledge upon residual capacity of corroded RC elements.

The iron oxides developed due to corrosion phenomenon on steel bars' surface occupy 4 to 6 times greater volume of the mass lost, generating tensile stresses in surrounding concrete with subsequent concrete cracking and spalling of the cover concrete. Hence, corrosion impairs the interface between steel and concrete and therefore affects the bond between them [21]. Bond is the imperative mechanism to denote the transfer of forces between reinforcement and surrounding concrete [22], which is mainly influenced by chemical adhesion, friction and mechanical interlock due to the presence of ribs.

Besides corrosion phenomenon, the aspects of steel – concrete adhesion depend to a high extent on numerous parameters related to both steel and concrete, namely steel bar geometry, concrete strength and confinement due to transverse reinforcement and concrete cover thickness [23]. The influence of compressive concrete strength on bond behavior of RC specimens has been studied by Abosrra et al. [24] and Zandi and Coronelli [25]. Yalciner et al. [26] conducted an experimental study on bond strength loss due to corrosion taken into account both the compressive strength of concrete f_c and the ratio of concrete cover thickness to nominal steel diameter c/D. Testing RC specimens without stirrups, Maslehuddin et al. [27] indicated that although a slight improvement of bond strength is demonstrated in low corrosion levels, sharp degradation pf bond strength is recorded as corrosion increases. Recently, experimental studies by Zandi et al. [28], Lin et al. [29] and Apostolopoulos and Koulouris [30] investigated both the significant role of stirrups spacing and corrosion carrying out eccentric pull out tests on RC elements with usual design values of concrete cover.

2. Corrosion of steel reinforcement

Steel reinforcement is the determinant factor of bearing capacity of reinforced concrete elements. However, in case of structures located in coastal regions (or marine environment), where high chloride contents are indicated, steel reinforcement degrades due to chloride-induced corrosion and subsequently leads to durability problems of the entire RC structures.

It is a common knowledge that steel reinforcement is initially protected by concrete cover and a passive layer on its surface. In particular, concrete cover thickness acts as a physical barrier between steel and corrosive environment, delaying the penetration and diffusion of corrosive factors through pores of concrete. At the same time, high alkalinity of concrete due to cement (pH ~ 12.5) results in protection of steel, forming a thin passive layer of ferric oxides on its surface. Chlorides reach the surface of concrete, enter the pore system either by diffusion (in stationary pore water), or by capillary suction of the surface water in which they are dissolved (or by combination of both transport mechanisms) [31]. It is assumed that there is an initiation period, until chloride ions reach the reinforcement, during which substances as water, chloride ions diffuse into concrete and reach the certain concentration necessary to trigger corrosion of the steel reinforcement [32]. This process has slow rate in nature, since it takes a long period of more than about 10 or 15 years, until the aggressive environmental agents reach the steel reinforcement, accompanied by reduction of concrete pH below 9, depassivate it and then corrosion initiates, **Figure 1**. When the passive protection breaks down then onset of steel corrosion (oxidation) takes place and gradually rust occurs on its surface.

Corrosion is an electrochemical phenomenon, in which the existence of an anode, a cathode, an electron pathway and electrolyte (ionic pathway) is required. The electrochemical reactions that occur during the corrosion process are presented as follows, in Eq. (1) and (2), [33]:

The anodic reaction (oxidation),

$$Fe \rightarrow Fe^{2+} + 2e^{-} \tag{1}$$

Since two electrons ($2e^-$) released in the above anodic reaction, there must be another reaction (cathodic reaction) in order to ensure the electrical neutrality on steel surface. This cathodic reaction, Eq. (2), consumes water and oxygen.

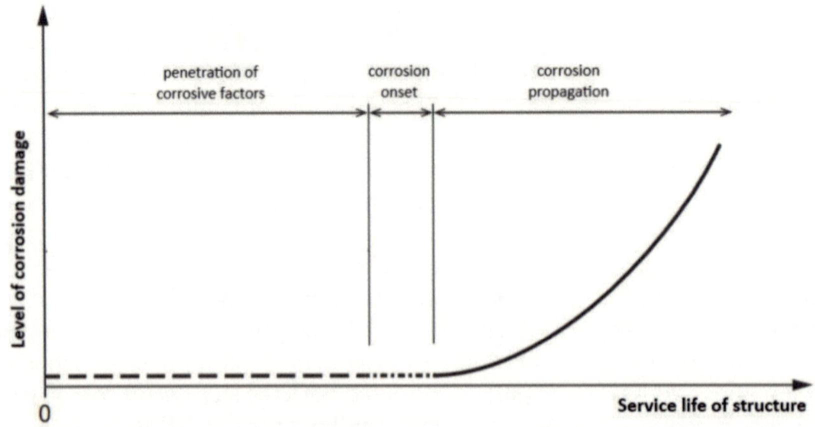

Figure 1.
Stages of corrosion in RC structures.

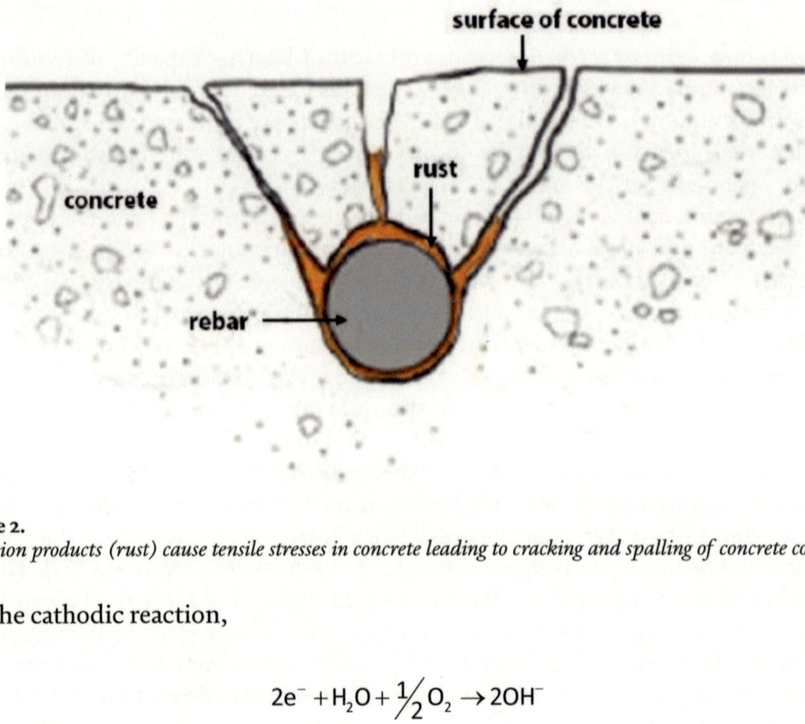

Figure 2.
Corrosion products (rust) cause tensile stresses in concrete leading to cracking and spalling of concrete cover.

The cathodic reaction,

$$2e^{-} + H_2O + \frac{1}{2}O_2 \rightarrow 2OH^{-} \tag{2}$$

The above two chemical equations contain the basic reactions at the first stages of corrosion. Then, as corrosion propagates, hydration reactions are followed, Eq. (3) - Eq. (5), so as to form hydrated ferric oxide (yield red rust) $Fe_2O_3 \bullet H_2O$:

$$Fe^{2+} + 2OH^{-} \rightarrow Fe(OH)_2 \tag{3}$$

where is $Fe(OH)_2$ ferrous hydroxide,

$$4Fe(OH)_2 + O_2 + 2H_2O \rightarrow 4Fe(OH)_3 \tag{4}$$

where $4Fe(OH)_3$ is ferric hydroxide and

$$2Fe(OH)_3 \rightarrow Fe_2O_3 \cdot H_2O + 2H_2O \quad (5)$$

where is $Fe_2O_3 \cdot H_2O$ hydrated ferric oxide (rust).

As shown in **Figure 2**, oxides (rust), which are formed due to corrosion on steel surface, occupy 2 to 6 times greater volume of the attacking mass [15], causing tensile stresses in surrounding concrete and, thereafter, leading to gradual concrete cracking development and spalling of the cover concrete. Hence, corrosion phenomenon affects significantly the steel reinforcement, reducing its initial cross-section, degrades its mechanical properties and bond between steel and concrete.

3. Laboratory testing

The capacity assessment of corroded reinforced concrete elements consists an engineering task of major importance, since the effects of steel corrosion have become more apparent. Nevertheless, the current international regulations and standards do not determine degradation rules of RC elements; thus, there is need to develop codes, taking into account the deterioration of materials and the subsequent reduction of structural capacity of RC members.

In order to study and quantify the consequences of corrosion, laboratory methods have been developed to simulate and accelerate the natural process. One of the most widespread accelerated corrosion techniques, used for the goals of many studies is impressed current density technique.

In this accelerated corrosion technique, the reinforcing steel bars (to be corroded) and stainless-steel bars are connected to the positive and the negative pole of a power supply, respectively, and are immersed in tanks, which are filled by a sodium chloride (NaCl) solution in content of 5% (by weight of water). In this way, an electric circuit is generated, since the reinforcing steel bars act as anode of circuit, the stainless-steel bars act as cathode and the NaCl solution is the electrolyte, which allows ions to flow into the circuit. Direct electric current is induced to the reinforcing steel bars through the power supply in order to accelerate the electrochemical reaction of corrosion [34]. This technique is in accordance with ASTM standards, although all the individual parameters needed to create a standard corrosive environment have not

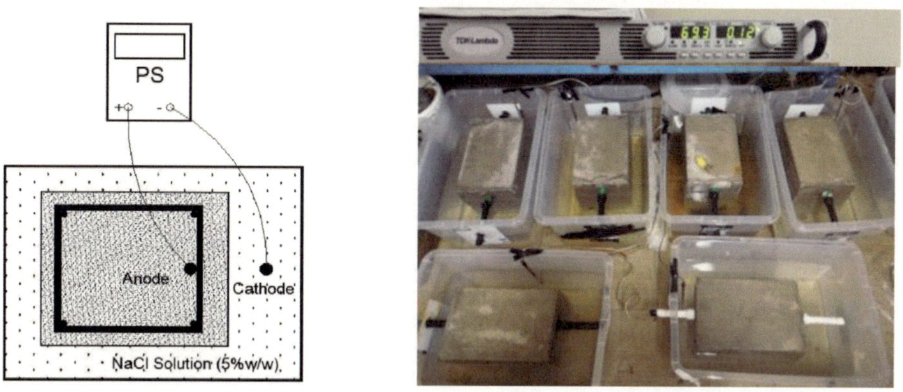

Figure 3.
Simplified procedure of accelerated electro-corrosion (left) - accelerated corrosion experiments on RC specimens via inducing current (right).

been determined yet. Nevertheless, the main advantage of this method is the ability to control the rate of corrosion, which usually varies due to changes in the resistivity, oxygen concentration, and temperature.

In **Figure 3** an indicative automatic system (found in Laboratory of Technology & Strength of Materials of University of Patras) is illustrated, which has been used for the performance of electrochemical corrosion tests. The specific system enables implementation of different corrosive conditions, in terms of impressed current density and ponding cyclic corrosion (wet/dry duration).

4. Differential aeration corrosion on steel reinforcement

In case of steel reinforcing bars, embedded in concrete, the corrosion damage is depicted in finite areas along their length, recording significant reduction of initial cross-section, as shown in **Figure 4**. Recent studies have indicated that the existence of voids and pores in concrete allows not only the penetration and diffusion of corrosive agents but also the higher oxygen concentration on steel bars' surface, consisting a favorable condition for differential aeration corrosion. The anodic dissolution rate depends solely upon the potential difference across the electrolyte-metal surface.

In order to further investigate and simulate the consequences of differential aeration corrosion, Apostolopoulos et al. [35] conducted accelerated corrosion experiments on bare reinforcing steel bars, taking into account different exposure to corrosion lengths. The results demonstrated that corrosion damage depends on the exposed to corrosion length, as short samples record higher mass loss percentages and more intense pitting, for the same corrosion duration. In particular, specimens with the short exposed to corrosion length demonstrated about 4 times greater percentage mass loss in contrast to specimens with the long exposed to corrosion length, for 300 h of accelerated corrosion time [35], which is due to Differential aeration corrosion.

In case of RC specimens of experimental study with weak concrete cover thickness, where bond forces are tested, differential aeration corrosion phenomena are detected on ribs. In particular, during the phase of concrete hardening, micro-cracks are recorded in the area of the edge of ribs, which lead to accelerate the penetration of aggressive agents to the steel reinforcement, starting from the damage from the ribs (as closer to the outer surface). Thus, ribs are more vulnerable to corrosion, resulting mainly in degradation of steel - concrete bond mechanism which leads to mechanical interlock's loss and rise of slippage. In that manner, corrosion primarily affects bond behavior of RC elements rather than their mechanical properties, even in low corrosion level. **Figure 5** illustrates intense pits on surface due to corrosion, which subsequently causes loss of ribs, especially on the outer half of the steel bar's circumference, adjacent to the external surface of RC specimens.

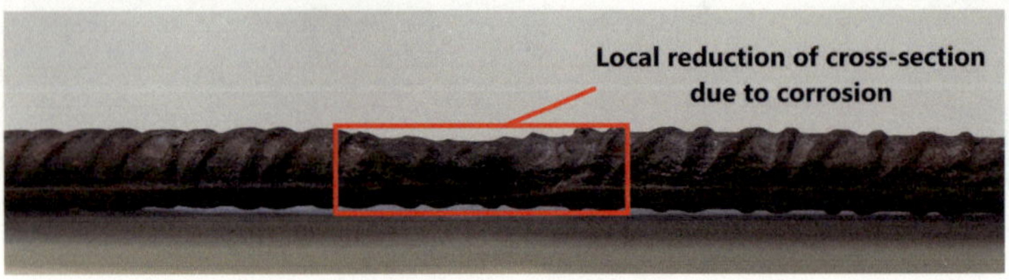

Figure 4.
Local reduction of cross-section due to corrosion.

Figure 5.
Corrosion damage on steel surface. Inner half of steel bar's circumference (up) - outer half of steel bar's circumference (down).

5. Corrosion effect on bond strength

Corrosion factor is responsible for reduction of the cross-sectional area of steel reinforcement, degrading the mechanical properties [9–14], and causing cracks and spalling of the cover concrete, which leads to deterioration of steel - concrete interface and, subsequently, to bond loss between them. In order to assess the structural integrity of corroded RC members, many researchers have studied and quantified the consequences of mechanical performance of steel reinforcement due to corrosion and proposed degradation material laws [15–16]. However, as it is abovementioned, bond behavior is primarily affected by corrosion, as the transfer of forces between reinforcement and surrounding concrete weakens. Thus, the reference to degradation of mechanical properties of steel reinforcement is not a central issue in this chapter.

Regarding the interface characteristics in RC elements, Model Code 2010 [36] recommends the calculation of ultimate bond strength f_{bd} in the non-corroded condition, as presented in Eq. (6), in which concrete and steel class, steel bar diameter and the contribution of confinement via cover thickness and transverse reinforcement are taken into account.

$$f_{bd} = (\alpha_2 + \alpha_3) \cdot f_{b,0} + \mathbf{2} \cdot p_{tr} \tag{6}$$

where $f_{b,0}$ is the basic bond strength, which depends on concrete and steel quality, bar geometry and the casting position of steel bar during concreting,

α_2 and α_3 are coefficients, which represent the influence of passive confinement from concrete cover thickness and from transverse reinforcement, respectively, in excess of their respective permissible minima, and.

p_{tr} is the mean compression stress perpendicular to the potential splitting failure surface at the ultimate limit state.

Prolonged exposure of reinforced concrete structures to a corrosive environment causes significant degradation problems in the steel-concrete bond mechanism, allowing relative slip to develop between steel and concrete and reducing the bond strength between them. Consequently, the bond loss effect due to the environmental action prevents the development of full bearing capacity of reinforced concrete elements until they behave as unreinforced members. From the abovementioned, it is obvious that bond mechanism is a main criterion in design of RC members. However, there is a gap in international codes, regarding the quantification of bond strength loss due to corrosion. In particular, even though range of values are proposed by Model Code 2010 for the estimation of reduced bond strength, considering corrosion penetration and surface crack, nevertheless the influence of stirrups spacing as well as the impact of non-uniform type of corrosion damage on steel bar's surface through pits, which is the most common in practice, are not determined yet. Due to this, there is area of research so as to establish degradation models, including the effect of both density of stirrups and local reduction of cross-section in maintenance of bond strength of corroded RC elements. Moreover, the term corrosion penetration, which is found in many regulatory texts, refers to a uniform circumference loss of the circular cross section due to mass loss, case of uniform corrosion which is practically non-existent in real RC structures exposed to chloride induced corrosion.

On this basis, many researchers have studied the effect of corrosion on bond between steel and concrete [24–30]. As Maslehuddin et al. [27] reported, bond strength of RC specimens without transverse reinforcement increases slightly in low corrosion degree; however, sharp bond loss takes place with the propagation of corrosion. These findings are in aggrement with the study of Auyeung [37], which demonstrated up to 80% bond strength loss due to only 2% reduction of steel cross-section. A more comprehensive study is presented by Lundgren [38] concerning the contribution of stirrups to the bond behavior of corroded RC elements. In real RC structures the estimation of corrosion penetration and mass loss of steel bars consists a difficult task, since steel bars are embedded in concrete and corrosion damage is non-uniform on steel's surface. However, cracks due to steel corrosion on concrete surface are visible and their width can be easily measured, as shown in **Figure 6**. For this reason, recent scientific studies tend to quantify the corrosion damage of the embedded steel reinforcement through the surface cracking width and, subsequently, estimate the local bond loss [10, 21, 24–30, 37–44]. An empirical correlation between the loss of steel bar's diameter and the average corrosion penetration has been proposed by Torres-Acosta et al. [39]. Moreover, studies of Andrade et al. [40] and Tahersamsi [10] link the surface cracking width

Figure 6.
Surface concrete cracking due to steel corrosion (left) -measuring of crack width on concrete (right).

with the corrosion damage of steel bar and the loss of bond strength. Recent experimental study of Lin et al. [29] investigated the influence of concrete cover thickness and stirrups on the occurrence of surface cracks and on the subsequent bond strength loss. Gathering various experimental data, a predictive model of bond strength loss as a function of surface concrete cracking has been suggested by Zhou et al. [41].

Based on the abovementioned, a broad and ongoing experimental research on corroded RC elements was conducted by Apostolopoulos and Koulouris [30], studying the influence of stirrups spacing (density) and concrete cover thickness on bond behavior of corroded RC specimens, in correlation with the surface cracking width due to corrosion. The results depict close correlation of bond strength with surface concrete cracking width.

The depassivation of protective layer on steel reinforcement leads to onset of corrosion, the propagation of which develops various range of surface concrete cracking along the reinforcing bars. As illustrated, firstly in the following **Figure 6** (Rigth) and thereafter in **Figure 7**, the surface cracking width varies depending on corrosion level, stirrups spacing and concrete cover thickness.

In the case of specimens with concrete cover of 25 mm, the values of average surface crack width followed a common path up to a low corrosion level of 3% (**Figure 7** Left). It is noteworthy that the specimens with dense stirrups (Φ8/60 mm) demonstrated initially higher values of cracking width, since corrosion potential was higher due to the high percentage of steel reinforcement; however, as corrosion degree increases the confinement provided by dense stirrups limited the progressive development of surface cracking. On the other hand, the specimens without stirrups) recorded initially limited range cracking, due to the low percentage of steel, whereas in higher corrosion levels the absence of confinement lead to rapid growth rate of cracking width.

Similar results were recorded in the case of specimens with concrete cover of 40 mm (**Figure 7** Right). More specific, specimens without stirrups and specimens with dense and quite dense stirrups (Φ8/60 mm and Φ8/120 mm respectively), for mass loss equal to 3%, depicted similar range of surface crack width. In contrast, specimens with stirrups Φ8/240 mm, for the same percentage of mass loss, recorded sudden and remarkable high range of surface crack. In this group pf specimens, namely with concrete cover thickness equal to 40 mm, the confined cross section is reduced in line to the specimens with concrete cover thickness equal to 25 mm. Hence, poor confinement level in conjuction with corrosion initiation impacts the uncontrolled propagation of surface cracking.

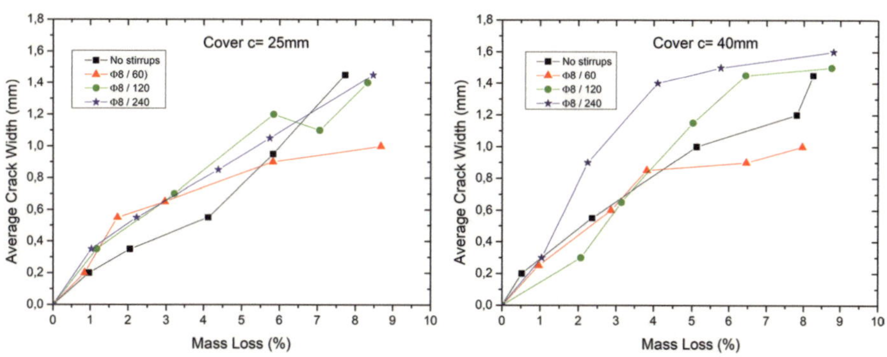

Figure 7.
Average crack width on concrete surface in function of percentage mass loss of steel bar. Cover thickness 25 mm (left) and 40 mm (right).

Nevertheless, with the evolution of corrosion, specimens with dense connectors, namely Φ8/60 mm, noted a significant decrement of surface cracking development. This particularly notable decrease, recorded in specimens with dense stirrups, was applied to both categories of specimens, one with concrete cover thickness of 25 mm and the other with concrete cover of 40 mm. More precisely, an average crack width equal to 1 mm has been recorded, for mass loss between 8.5% and 9%, respectively.

It is obvious that surface crack width is the outcome of corrosion damage of steel reinforcement; surface cracking is directly linked to various parameters beginning with the existence and amount of transverse reinforcement and cover concrete thickness. For both groups of specimens with different concrete cover (25 mm and 40 mm, respectively), the presence of dense stirrups (Φ8/60 mm) is preceded with a remarkable limitation of the surface cracking evolution to a width threshold of 0.90 mm, corresponding value to the abovementioned average mass loss of 8.5%–9.0%.

In order to investigate the bond behavior, pull out tests of uncorroded and corroded RC specimens conducted, the results of which confirmed that, in both cases of concrete cover thickness of 25 mm and 40 mm, the increase of the average range of surface cracking brought a dramatic decrease of bond strength between concrete and steel bar, **Figure 8**. Moreover, obtained by non-linear regression analysis, exponential predictive models of bond strength loss due to corrosion of steel reinforcement were given, derived from the correlation of bond strength loss of corroded specimens and surface cracking of concrete. The functions of predictive models are as follows, Eq. (7):

$$\frac{c_b^{cor}}{c_b^{uncor}} = e^{-A \cdot c_w} \tag{7}$$

where A is a parameter that depends on the concrete cover (c) and the amount of transverse reinforcement (stirrups spacing or absence of stirrups). The values of parameter A for each predictive model, and the corresponding values of the R^2 coefficient, are presented in the following **Table 1**. As shown by the predictive curves, stirrups spacing is main influencing factor of bond strength degradation due to corrosion. Hence, there is need to determine specific models of bond loss in order to enhance the current technical codes.

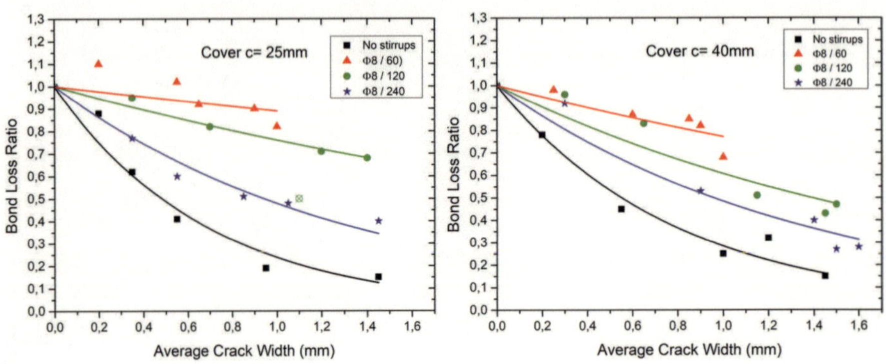

Figure 8.
Predictive models of bond strength loss correlated to the average surface crack width. Cover thickness 25 mm (left) and 40 mm (right).

Cover (mm)		No Stirrups	Φ8/240	Φ8/120	Φ8/60
25	A	1.435	0.736	0.274	0.117
	R^2 (%)	96.2	96.5	97.7	45.9
40	A	1.257	0.724	0.499	0.260
	R^2 (%)	97.5	96.0	91.9	80.1

Table 1.
Parameters (by regression analysis) for the exponential predictive model of bond strength loss.

Among specimens of the same concrete cover, it is clear that densification of transverse reinforcement slows down the progression of bond loss. In particular, specimens with concrete cover equal to 25 mm and dense stirrups Φ8/60 mm, the bond strength performance, even though its initial increase up to a threshold of 0.60 mm crack width, remained stable as in the case of non-corroded specimens. Moreover, among specimens of similar range of cracking, specimens without stirrups recorded a decline of bond strength performance equal to 57%, whereas specimens with wide stirrups (Φ8/240 mm) and quite dense stirrups (Φ8/120 mm) recorded a decrease of bond strength performance equal to 35% and 15%, respsectively.

The wide stirrups spacing (Φ8/240 mm), in case of 25 mm cover thickness, degrades bond strength performance, leading to a residual bond strength equal to 40% of the non-corroded bond strength, corresponding to a crack width of 1.45 mm, whereas the absence of stirrups (specimens without stirrups) recorded bond loss equal to 16% of non-corroded value. The abovementioned outcomes transposed to former practices, as in the existing building stock, the use of transverse reinforcement accounts for four pieces per linear meter, i.e. Φ8/250 mm. Thus, loss of bond strength seems to be inevitable.

In the case of cover thickness of 40 mm, the absence of stirrups deteriorates rapidly bond strength contrary to the dense fitting of stirrups (Φ8/60 mm) where this benefits, thereby delaying bond strength degradation.

The bond strength between steel and concrete demonstrates a denoting drop when increasing the range of surface cracking; it follows from the assessment of both cases of concrete cover that as the range of surface cracking raises, the threshold of bond strength performance reduces. These results come in good agreement with results of former studies, to name Lin et al. [29], Fischer and Ozbolt [42], Almusallam et al. [43] and Rodriguez et al. [44].

Given the tendency to approach the issue of bond strength between concrete and steel, exponential predictive models were developed in order to link the bond strength loss of corroded specimens to the average width of concrete surface cracking. The exponential model introduces an adequate approach of bond strength loss and comes to agreement with previous studies [43, 44] as corrosion evolves. Hence, so as to assess the bond loss, besides the traditional method of chlorides' measurement, surface cracking measurement occurs as an emerging methodology. Notwithstanding the efforts of the scientific community to correlate experimental results of current literature, the issue of dispersion is pertinent due to several parameters, to cover thickness, concrete class, nominal diameter of steel reinforcement, presence of stirrups. It is also noteworthy that, in existing experimental literature, exponential models are proposed as predictive models of bond behavior of corroded RC elements, regardless of the differences denoted in all parameters to-be-tested in comparative studies.

Extending the investigation of bond behavior of corroded RC members and focusing on values of maximum pull-out force and, subsequently, bond strength, and not on bond loss, the role of stirrups spacing is more highlighted, where bonding between steel and concrete degrades due to corrosion and the developed maximum pull-out force drops. The usage of dense transverse reinforcement contributes to bond behavior, not only reducing the bond loss rate, but leading to greater values of maximum pull-out force due to confinement, **Figure 9**.

In an effort to estimate that influence of stirrups spacing on non-corroded condition, the presence of wide stirrups (Φ8/240 mm) present maximum bond strength equal to 7.04 MPa, whereas quite dense (Φ8/120 mm) and dense stirrups (Φ8/60 mm) equal to 9.10 MPa and 9.53 MPa, respectively. The percentages attributing the increase of bond strength against specimens without stirrups are 35.9%, 75.6% and 84%, respectively. It is noteworthy that, quite dense stirrups spacing (Φ8/120 mm) result in sufficient bond strength levels, whereas further densification leads to minor increase of bond strength, nevertheless delays the bond loss. Extrapolating the abovementioned on real RC structures, stirrups' densification above a certain threshold, is considered inappropriate since it could lead on the one hand to substantial increase of costs and on the other hand to rapid rise of corrosion rate, due to potential increase.

Moreover, bond stress - relative slip curves are exported from pull-out tests for each corrosion level and for each category of stirrups spacing, respectively. Typical curves of uncorroded and corroded specimens are shown in **Figure 10**.

The harmful influence of corrosion phenomenon on bond behavior between steel and concrete can be seen examining the bond stress - slip curves of **Figure 10**, and in particular group of specimens without stirrups. Uncorroded specimens - where no surface cracking had been initially observed - showed a quasi linear relationship between bond stresses and relative slip till the point of bond strength's development. During this phase, cracks were occurred, parallel to the axis of steel bar, and gradually were developed due to the radial stresses, which are transferred from steel to concrete. After the development of bond strength, sharp decline of bond stresses and complete bond loss followed. In the case of corroded specimens (without

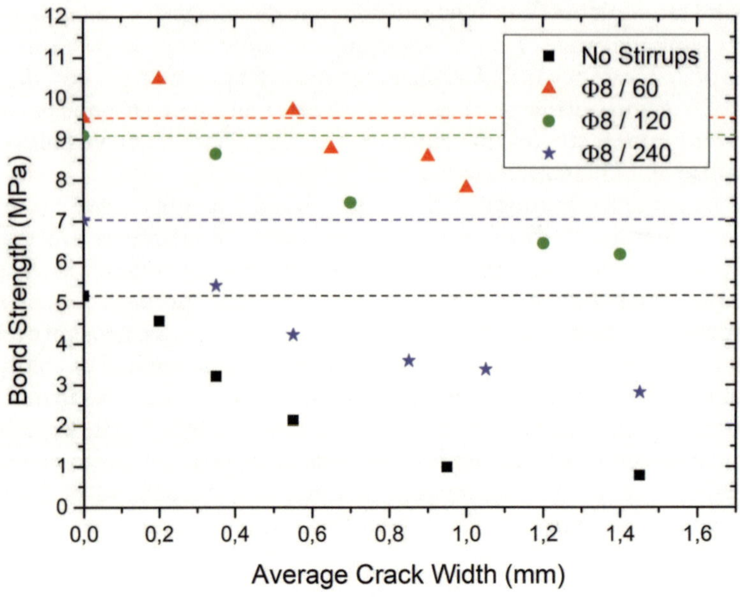

Figure 9.
Bond strength values in function with average crack width.

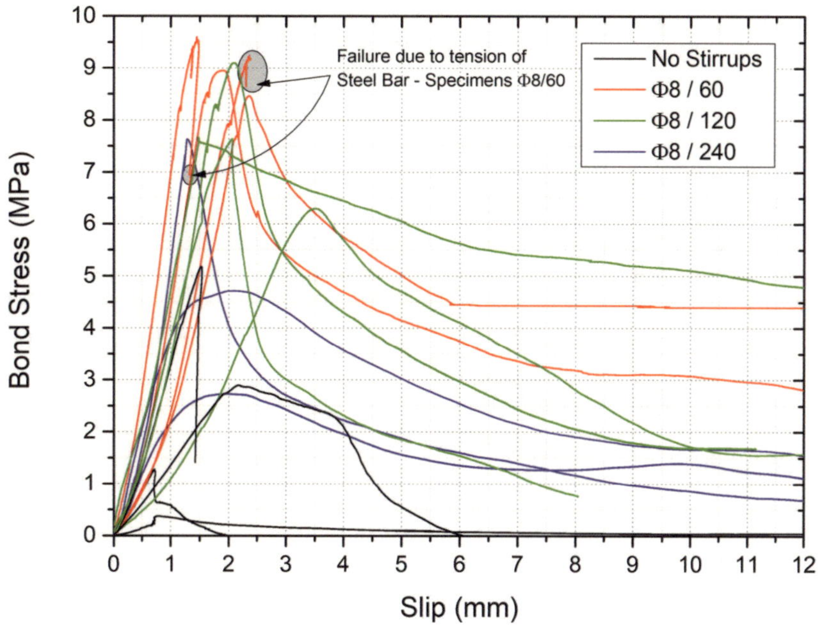

Figure 10.
Bond stress - slip curves obtained from pull out tests.

stirrups) intense surface concrete cracking was recorded, due to steel corrosion, even in low mass loss levels. As a result of this cracking, bond behavior degrades dramatically, as confirmed by the corresponding curves of **Figure 10**, since both the steel-concrete interface is damaged by the corrosion oxides and the confinement level is deteriorated by the cracks in concrete cover.

On the other hand, as shown by the groups of specimens with stirrups, surface cracking and degree of bond loss is strongly correlated with stirrups spacing. In particular, specimens with quite wide stirrups (Φ8/240 mm) showed that transverse reinforcement has a positive impact on bond behavior in uncorroded condition, as greater values of bond stress indicated. However, when corrosion occurs, accompanied by surface concrete cracking, bond resistance degradates as reflected in responding curves, where bond strength reduces significantly, and subsequently low values of residual bond stress are recorded.

Greater contribution of stirrups to bond behavior was noticed on specimens with quite dense stirrups spacing (Φ8 /120 mm), where even higher bond strength values are indicated, and while bond behavior degrades due to corrosion damage nevertheless does not demonstrate a massive drop mainly due to confinement. In addition, in the case of wide stirrups (Φ8 / 60 mm), full use of bond behavior, even after corrosion damage, occurs, with residual bond stress recorded after the peak of bond strength. It is also noteworthy that specimens of this category had finally ended due to failure under tension. Thus, that densification of stirrups make full use of bearing capacity of steel reinforcing bars and subsequently of RC elements. From the aforementioned, it becomes obvious that the use of dense stirrups (Φ8/60 mm) leads to full bonding between steel and concrete, in uncorroded and corroded conditions respectively, marking both high bond strength and high residual bond stress after the ultimate pull out force.

To conclude, existence of stirrups contributes to bond behavior, and subsequently, greater density of stirrups affects both the reduction of bond loss' degradation rate due to steel corrosion, as well as the increasing of bond stresses values due to greater confinement. The use of wide stirrups (Φ8 / 240 mm) enhances the bond strength, in uncorroded conditions, about 35.9% in comparison with group of specimens without stirrups. However, when corrosion occurs, the subsequent surface cracking degrades significantly the confinement and the bond between steel and concrete, about 60% reduction against of uncorroded specimens with Φ8 /240 mm. Specimens with quite dense stirrups (Φ8 /120 mm) indicated higher bond strength values, and while bond behavior degrades due to corrosion damage nevertheless does not demonstrate a massive drop mainly due to confinement, about 32%. Finally, Dense stirrups spacing, specimens with Φ8/60 mm, ensure high level of bond behavior, either in terms of bond strength or of residual bond stress, both in uncorroded and corroded conditions. Furthermore, stirrups spacing of 60 mm results in full anchorage of steel reinforcing bars and make full use of their bearing capacity.

6. Conclusions

The present chapter presents an extensive and ongoing experimental research, which was conducted in Laboratory of Technology and Strength of Materials, in University of Patras, and comes to agreement with corresponding results of other scientific studies. The effect of steel corrosion on bond loss between steel and concrete was deeply investigated, including influencing parameters such as concrete cover thickness, density of stirrups and surface concrete cracking. The results of this research, the following outcomes were obtained:

- The width of surface cracking on concrete due to corrosion of steel reinforcement is closely related both to the cover thickness and to the amount of stirrups or their absence in RC element.

- Existence of stirrups contributes to bond behavior, and subsequently, greater density of stirrups affects both the reduction of bond loss' degradation rate due to steel corrosion, as well as the increasing of bond stresses values due to greater confinement.

- The development of surface cracking in concrete is associated with an exponential reduction of bonding forces.

- Based on the fact that presence of dense connectors, Φ8/60 mm, was accompanied by a clear limitation of the surface cracking development, it appears that the densification of stirrups (through the confinement) contributes positively to maintaining bond between steel reinforcement and concrete.

- In conclusion, there is a need for further improvement and strengthening of existing technical codes, introducing predictive models of bond loss in function with corrosion damage, cover thickness, surface concrete cracking and density of stirrups.

Author details

Charis Apostolopoulos and Konstantinos Koulouris*
Department of Mechanical Engineering and Aeronautics, University of Patras, Patras, Greece

*Address all correspondence to: kkoulouris@upnet.gr

References

[1] G. Koch, J. Varney, N. Thompson et al., International Measures of Prevention, Application, and Economics of Corrosion Technologies Study, NACE International, Houston, TX, USA, 2016.

[2] Hou B, Li X, Ma X, et al. The cost of corrosion in China. NPJ Materials Degradation. 2017;**1**(1):1-10

[3] Liu Y et al. Modeling the time-to-corrosion cracking in chloride contaminated reinforced concrete structures. ACI Materials Journal. 1998;**95**(6):675-681

[4] Angst U, Elsener B, Larsen CK, Vennesland Ø. Critical chloride content in reinforced concrete—A review. Cement and Concrete Research. 2009;**39**(12):1122-1138

[5] Babaee M, Castel A. Chloride diffusivity, chloride threshold, and corrosion initiation in reinforced alkali-activated mortars: Role of calcium, alkali, and silicate content. Cement and Concrete Research. 2018;**111**:56-71

[6] Cao Y, Gehlen C, Angst U, Wang L, Wang Z, Yao Y. Critical chloride content in reinforced concrete — An updated review considering Chinese experience. Cement and Concrete Research. 2019;**117**:58-68

[7] Richardson MG. Fundamentals of Durable Reinforced Concrete. London: Spon Press; 2002

[8] BSI (British Standards Institution). Concrete. Part 1: Specification, Performance, Production and Conformity." EN 206-1. London; 2000

[9] Fernandez I, Bairán JM, Marí AR. Mechanical model to evaluate steel reinforcement corrosion effects on σ–ε and fatigue curves. Experimental calibration and validation. Engineering Structures. 2016;**118**:320-333

[10] Tahershamsi M, Fernandez I, Lundgren K, Zandi K. Investigating correlations between crack width, corrosion level and anchorage capacity. Structure and Infrastructure Engineering. 2017;**13**(10):1294-1307. DOI: 10.1080/15732479.2016.1263673

[11] Zhang W, Song X, Gu X, Li S. Tensile and fatigue behavior of corroded rebars. Construction and Building Materials. 2012;**34**:409-417

[12] Sun X, Kong H, Wang H, Zhang Z. Evaluation of corrosion characteristics and corrosion effects on the mechanical properties of reinforcing steel bars based on three-dimensional scanning. Corrosion Science. 2018;**142**:284-294

[13] Andisheh K, Scott A, Palermo A, Clucas D. Influence of chloride corrosion on the effective mechanical properties of steel reinforcement. Structure and Infrastructure Engineering. 2019;**15**(8):1036-1048

[14] Gu X, Guo H, Zhou B, Zhang W, Jiang C. Corrosion non-uniformity of steel bars and reliability of corroded RC beams. Engineering Structures. 2018;**167**:188-202

[15] Apostolopoulos C, Papadakis VG. Consequences of steel corrosion on the ductility properties of reinforcement bar. Construction and Building Materials. 2008;**22**(12):2316-2324

[16] Apostolopoulos C, Kappatos V. Tensile properties of corroded embedded steel bars B500c in concrete. International Journal of Structural Integrity. 2013;**4**(2):275-294

[17] Fernandez I, Herrador MF, Marí AR, et al. Structural effects of steel reinforcement corrosion on statically indeterminate reinforced concrete members. Materials and Structures.

2016;**49**:4959-4973. DOI: https://doi.org/10.1617/s11527-016-0836-2

[18] Zandi K. Structrural behavior of deteriorated concrete structures [PhD thesis]. Gothenburg, Sweden: Chalmers University of technology; 2010

[19] Zhu W, François R. Prediction of the residual load-bearing capacity of naturally corroded beams using the variability of tension behaviour of corroded steel bars. Structure and Infrastructure Engineering. 2016;**12**(2):143-158. DOI: 10.1080/15732479.2014.996165

[20] Kashani MM, Maddocks JR, Dizaj EA. Residual capacity of corroded reinforced concrete bridge components: State-of-the-art review. Journal of Bridge Engineering. 2019;**24**(7)

[21] Fu X, Chung DDL. Effect of corrosion on the bond between concrete and steel rebar. Cement and Concrete Research. 1997;**27**(12):1811-1815

[22] Tepfers, R., Z. Achillides, A. Azizinamini, G. Balázs, Agniezka Bigaj-van-Vliet, J. Cabrera, J. Cairns, E. Cosenza, J. D. Uijl, R. Eligehausen, B. Engström, L. Erdélyi, P. Gambarova, J. Jirsa, S. Lane, R. León, J. Magnússon, U. Mayer, S. Mccabe, C. Modena, J. Modniks, T. Mottram, K. Noghabai, Koiji Otsuka, J. Ožbolt, Stavrola J. Pantazopoulou, K. Pilakoutas, G. Plizzari, R. Realfonzo, Jesós Rodriguez, G. Rosati, G. Russo, S. Russo, H. Shima, Christiano Schumm, L. Taerwe, V. Tamuzs, T. Ueda, L. Vandewalle and lisabeth Vintzileou. "Fib Bulletin 10. Bond of Reinforcement in Concrete." (2000).

[23] Li CQ, Zheng JJ. Propagation of reinforcement corrosion in concrete and its effects on structural deterioration. Magazine of Concrete Research. 2005;**57**(5):261-271

[24] Abosrra L, Ashour AF, Youseffi M. Corrosion of steel reinforcement in concrete of different compressive strengths. Construction and Building Materials. 2011;**25**(10):3915-3925

[25] Zandi K, Coronelli D. Anchorage capacity of corroded reinforcement: Eccentric pull-out tests on beam-end specimens. In: Report No. 2010-06, Department of Civil and Environmental Engineering. Goteborg, Sweden: Chalmers University of Technology; 2010

[26] Yalciner H, Eren O, Sensoy S. An experimental study on the bond strength between reinforcement bars and concrete as a function of concrete cover, strength and corrosion level. Cement and Concrete Research. 2012;**42**(5):643-655

[27] Maslehuddin M, Allam I, Al-Sulaimani G, Al-Mana A, Abduljauwad S. Effect of rusting of reinforcing steel on its mechanical properties and bond with concrete. ACI Materials Journal. 1990;**87**(5):496-502

[28] Zandi K, Coronelli D, Lundgren K. Bond capacity of severely corroded bars with corroded stirrups. Magazine of Concrete Research. December 2011;**63**(12):953-968. DOI: https://doi.org/10.1680/macr.10.00200

[29] Lin H, Zhao Y, Ozbolt J, Hans-Wolf R. Bond strength evaluation of corroded steel bars via the surface crack width induced by reinforcement corrosion. Engineering Structures. 2017;**152**:506-522

[30] Apostolopoulos C, Koulouris K, Apostolopoulos A. Correlation of surface cracks of concrete due to corrosion and bond strength (between steel Bar and concrete). Advances in Civil Engineering;**2019**:12. DOI: https://doi.org/10.1155/2019/3438743

[31] Ozbolt J, Balabanic G, Periskic G, Kuster M. Modelling the effect of damage on transport processes in

concrete. Construction and Building Materials. 2010;**24**:1638-1648

[32] Apostolopoulos C et al. Chloride-induced corrosion of steel reinforcement – Mechanical performance and pit depth analysis. Construction and Building Materials. 2013;**38**:139-146

[33] Broomfield JP. Corrosion of Steel in Concrete: Understanding, Investigation and Repair. London: E & FN Spon; 1997

[34] Sh A. Techniques for inducing accelerated corrosion of steel in concrete. Arabian Journal for Science and Engineering. 2009;**34**(2C):95-104

[35] Apostolopoulos C, Koulouris K, Basdeki M. Damage parameters of rebars in marine environment and fatigue life. In: Rilem SMSS Conference 2019.

[36] Model Code EBCEB-FIP. 2010. fib Model Code Concr. Struct. 2010. Lausanne, Switzerland. 2013:152-189

[37] Auyeung Y, Balaguru P, Chung L. Bond behavior of corroded reinforcement bars. ACI Materials Journal. 2000;**97**:214-220

[38] Lundgren K. Effect of corrosion on the bond between steel and concrete: An overview. Magazine of Concrete Research. 2007;**59**(6):447-461

[39] Torres-Acosta AA, Navarro-Gutierrez S, Teran-Guillen J. Residual flexure capacity of corroded reinforced concrete beams. Engineering Structures. 2007;**29**(6):1145-1152

[40] Andrade C, Cesetti A, Mancini G, Tondolo F. Estimating corrosion attack in reinforced concrete by means of crack opening. Structural Concrete. 2016;**17**(4):533-540

[41] Zhou HJ, Zhou YF, Xu YN, Lin ZY, Xing F. Regression analysis of bond parameters between corroded rebar and concrete based on reposted test data. International Journal of Corrosion. 2018, 2018:18 Article ID 5309243

[42] Fischer C, Oˇzbolt J. An appropriate indicator for bond strength degradation due to reinforcement corrosion. In: Proceedings of the 8th International Conference on Fracture Mechanics of Concrete and Concrete Structures (FraMCoS). Toledo, Spain; March 2013. pp. 1828-1835

[43] Almusallam A, Al-Gahtani A, Aziz A, et al. Effect of reinforcement corrosion on bond strength. Construction and Building Materials. 1996;**10**(2):123-129

[44] Rodriguez J, Ortega LM, Garda AM. Assessment of structural elements with corroded reinforcement. In: Proceedings of International Conference Corrosion and Corrosion Protection of Steel in Concrete. Vol. 1. Sheffield, England: University of Sheffield; July 1994. pp. 172-185

Chapter 4

Risk Assessment in the Monitoring of Works

M. Rosário Oliveira

Abstract

The purpose of this chapter is to present a methodology for developing Control, Measurement and Monitoring Plans. It aims to apply risk-based thinking associated with the works control plan. The failures and rework of the works must not be accepted as inevitable or even as certainties. They must be considered permanent challenges to their management. The importance of using risk assessment techniques in the planning and control of the production activities of the works is evident. Control, measurement and monitoring process should provide the assessment of risks and failures, should demonstrate technical compliance of the work, and should improve operational efficiency. Thus, it is important to define a methodology for the preparation of the Control, Measurement and Monitoring Plan (PCMM), to be implemented in the execution of the works, in order to ensure the conformity of the works with its technical and regulatory requirements. It must establish which the trials and control inspections, its acceptance criteria, its purposes, frequencies and responsible and it must also identify and assess its risks.

Keywords: Control, Measurement and Monitoring Plan, analysis and risk assessment, building works

1. Introduction

Building works productivity is not only improved with more works and more companies. It improves with more competitive companies and with better organizational and technological capacity.

The work of the Construction is developed within a growing and demanding context where rigor and competence in production management are necessary conditions for the provision of the best service and product, being also essentials to maximize the profitability of the works [1].

For building companies, the logical choice to ensure competitive advantage with the rest of industry requires the use of new productivity tools and work production control methods. In many manufacturing industries, production processes have been modified with the implementation of systems that limit the existence of the failures and reworks along the production flow. These industries are confined to factories and can implement efficient monitoring systems that define any process accurately and subsequently monitor their implementation [2, 3].

Defects and rework should not be accepted as inevitable or even as certainties but considered as a permanent challenge to the management of the works, being important to use risk assessment techniques there its planning and control [2].

The "Operational Planning and Control" requirement specified in ISO 9001: 2015 indicates that organizations must plan, implement and manage the processes necessary for the supply of the product and service provision (Works) to ensure compliance with customer requirements (Owners) [4, 5].

Control, measurement and monitoring process of the works should provide the specific actions to address the risks and opportunities and achieve the objectives specified in their planning [4].

Thus, it is necessary to establish Plans of Control, Measurement and Monitoring Plans (PCMM), assess risk and define actions for its treatment, and implement the control of operational processes in accordance with the defined criteria. Plans of Control, Monitoring and Measurement (PCMM) are required to: i) demonstrate the technical compliance of the Work; ii) continuously improve operational effectiveness [2].

2. Risk assessment and plans of control

Risk assessment is an integral part to the various process of the works, aiming at prevention and its resilience.

To understand the risk assessment is necessary to know the definition adopted for "*risk*" in ISO Guide 73 (*Risk management – Vocabulary - Guidelines for use in standards*) and ISO 31000 (*Risk management - Principles and guidelines*). In according to these standards' "*risk*" is defined as the "*effect of uncertainty on objectives*" [6, 7].

This definition gives the possibility of we considering the risk as a threat or an opportunity. However, it is not gives clue as to how to quantify the risk. For this purpose, we must use the definition of the risk as being the combination of the probability of the occurrence an event (Likelihood or Frequency) and its consequences (Severity), something that is referred to in complementary notes these standards. The risk assessment compares the results of risk analysis with risk criteria (frequency and severity criteria) to determine whether the risk is acceptable or tolerable. It requires the identification and analysis of the events (occurrence or change of a set of circumstances) and to determine the risk level. The risk level is a function of its consequence (or Severity) with its likelihood (or Probability) and measures the magnitude of the risk. The risk valuation criteria are references in respect of which the significance of the risk is assessed [8].

The events that influence the results of processes under analysis can be identified and classified between risks and opportunities. The opportunities are directed to the organization's strategic planning processes and the risks are analyzed by quantifying the probability of occurrence and the severity its effects, to determine its level and the actions of mitigation [4].

Risk assessment may be made at difference degrees of depth and detail, we using various techniques ranging from simple to complex. We must use the risk criteria consistent with the scope of the process under analysis, as well as the technique and the assessment results. Likelihood/Consequence Matrix (LCM) and Failure Mode and Effects Analysis (FMEA) are two of several techniques of great application in risk assessment.

Likelihood/Consequence Matriz (LCM) is a technique that combines the probability of the event under analysis with its effects, to define a qualification of the level of risk. The form of the matrix and the definitions that apply to it depend on the context in which it is used. This technique is used to classify risks and their sources, and to identify your treatments. It can be used in situations where there is not enough data for a detailed analysis or when the situation does not justify the time and effort for a more quantitative analysis. It is relatively easy to use, and it provides quick ordering of risks at different levels of significance. Adequate scales of the likelihood and consequences criteria and the definition of risk matrix are the inputs essential the risk assessment process. The likelihood criteria scale (Probability) should cover the relevant domain for the case in analysis. The consequence criteria scale (Severity) should cover the range of different types of the consequences to be considered, from the maximum plausible consequence to the smallest plausible consequence to be considered. All scales can have any number of the levels, the most common being the scales of 3, 4 or 5 levels [7].

To order the risks, the consequence descriptor (Severity) that best suits the situation is chosen first, then the probability (Probability) of occurrence of these consequences is then defined. The risk level defined in the LCM may be associated with a decision rule, such as, for example, treating or not treating the risk.

Failure Mode and Effects Analysis (FMEA) is a technique for analyzing the reliability of the products, systems or processes. It is used to identify modes in which components of products, systems or processes may to fail to performance their functions. There are several FMEA applications: design FMEA which is used for components and products; system FMEA which is used for systems and process FMEA which is used for manufacturing processes and procedures and assembly. FMEA is also used in risk assessment and this requires detailed information on the phases of the case under study, to permit a significative analysis its failure modes. To perform a FMEA is fundamental the experience of the evaluators, the knowledge of the history of the failures and the causes, of the decision criteria and/or acceptance of the specific risks, and of the steps of the case under study [9].

Severity, Probability and Detection indices are the inputs for FMEA. Their scales must be adequate to the consequences and the likelihood of the events that combined define the risk matrix. Additionally, the level of risk combined with the failure detection index determines the Risk Priority Number (RPN). The scales these three criteria can have any number of levels, the most common being scales of 3, 5 or 10 levels [7, 8, 10–13].

To order the risks identifies, the consequence descriptor is chosen first, which best adapts to the situation, then the probability of occurrence of that consequence is defined. With the third detection descriptor, the Risk Priority Number (RPN) is defined. RPN is used to prioritize the of risk mitigation actions.

If we accept that all results of the building works processes are subject to uncertainty, then we can conclude that there is need a risk assessment for each of these. So, we can find the risk assessment in the reception of materials and the control of work in progress, in short, in all critical processes of the works, that it can ensure your technical and regulatory conformity.

According to ISO 9001 standard, production and service provision processes should be implemented under controlled conditions. This determines that the operational processes of the works are implemented a controlled mode, before, during and after its completion, particularly all its critical activities. Among other requirements, this condition includes the implementation of monitoring and measurement activities, in adequate steps, to verify if the criteria of the process or its outputs and the criteria for acceptance of the product and services were satisfied.

With this aim, it is essential to establish Plans of Control, Measurement and Monitoring (PCMM), assess the risk and define actions for its treatment, and implement the control of operational processes in conformity with the defined criteria. The PCMM is the document that specifies which are the trials and control inspections, the purposes, the acceptance criteria, the frequency, those responsible for the monitoring, and the records of the results obtained, in order to retain the objective evidence to the satisfaction of technical and regulatory requirements of the Work [2, 14].

3. PCCM form with risk assessment

This heading, on context of the building works processes, a methodology is proposed for elaboration of the PCMM, following the approach of risk-based thinking. It applies to operational processes considering the most critical work activities.

According to [2] the steps to be followed in the preparation of the PCMM require the definition of the: critical works activities that need to be controlled; inspections and tests to be performed in each critical activity; acceptance, frequency and sampling criteria of the inspections and tests; those responsible for control, measurement and monitoring; records of the results; risk assessment and its effects; and corrective and preventive actions to be implemented. **Figure 1** shows the flowchart of the methodology for preparing the PCMM.

3.1 Risk assessment with Likelihood/Consequence Matriz (LCM)

Table 1 shows the template PCMM with LCM which takes the form of a matrix of the columns and rows whose contents are explained below.

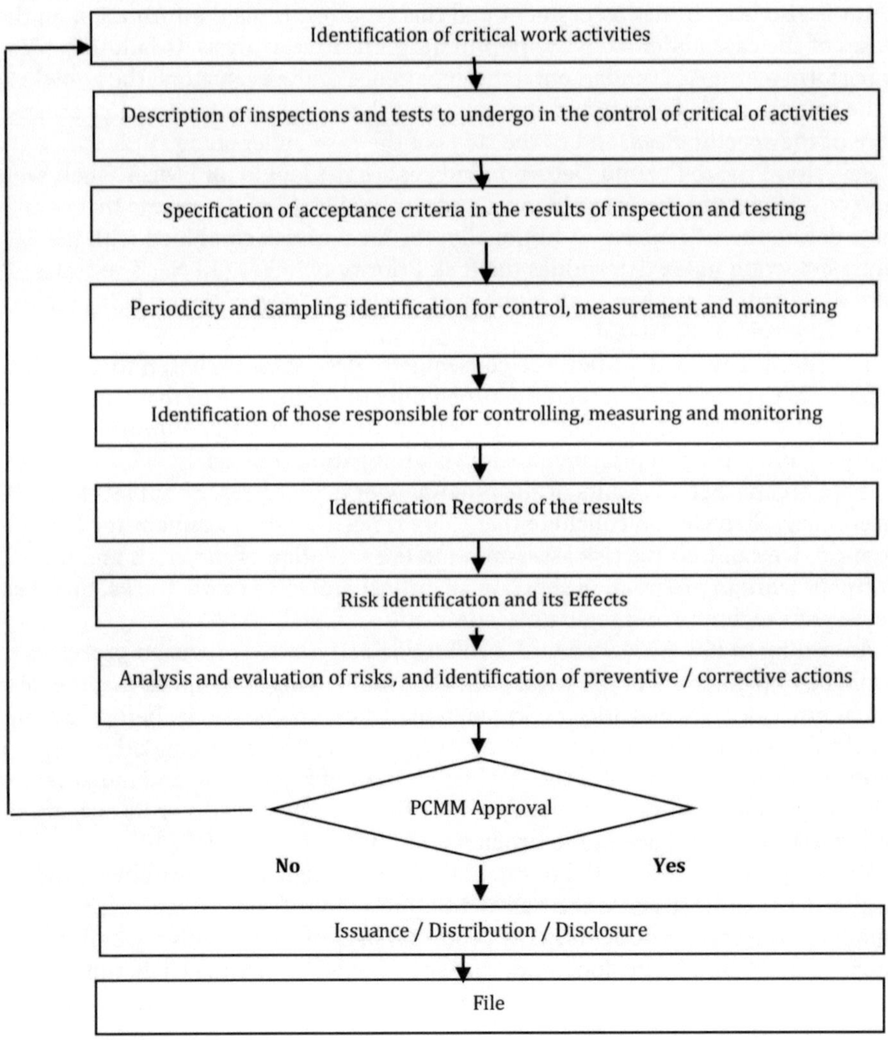

Figure 1.
Flowchart for preparing PCMM [2].

DOI: http://dx.doi.org/10.5772/intechopen.93957

Logo					PCMM - Plan Control, Measurement and Monitoring									Revision:	
														Page:	1/1
					Process/Work:										
Item (1)	Activities (1)	Inspection/ Testing (2)	Purposes (3)	Acceptance criteria (4)	Frequency/ Sampling (5)	Responsible (6)	Re cords (7)	Risks (8)	Description (9)	Effects (10)	Risk Assessment (11)			Controlled Risks? (12)	Actions (13)
											P	S	RN		

Logo						PCMM - Plan Control, Measurement and Monitoring								Revision:	
														Page:	1/1
						Process/Work:									
Item (1)	Activities (1)	Inspection/ Testing (2)	Purposes (3)	Acceptance criteria (4)	Frequency/ Sampling (5)	Responsible (6)	Re cords (7)	Risks (8)	Description (9)	Effects (10)	Risk Assessment (11)			Controlled Risks? (12)	Actions (13)
											P	S	RN		
Prepared by:		Date:		Verified by:		Date:				Responsible:				Date:	
		_/_/_				–/–/–								–/–/–	

Table 1.
PCMM template with LCM [2].

Using the LCM technique for risk assessment in the preparation of the PCMM, in according [2] the following guidelines must be used:

Columns (1) of the PCMM is identified the critical activities of the work that will be monitored. The critical activities are those ensure the technical conformity of the work with your design.

In the columns (2-3) are used to indicate the kind of the inspections and tests to be applied in quality control, their purpose and that is to be measured to monitor critical activity in analysis. The column (4) is used to propose which normative and regulatory references and criteria for acceptance that are used to analyze the results obtained in the inspections and in the tests carried out.

At each inspection/test, in the columns (5-7) are indicated their sampling frequency, who is responsible for carrying out the control and which record to use to compile their results. These items are intended to ensure the systematic control, measurement and monitoring in each critical activity.

Risk assessment is carried out at each stage of control, measurement and monitoring. Thus, for each critical activity is identified the events whose outputs may not be what expected.

The identification of risk and its effects is done on the columns (8-10) of the PCMM. For each critical activity and its control, the Risk is identified, its description is made and its effect is characterized. In this way, the event associated with it will be featured and you can review the respective controls, measurements and monitoring, especially in cases where it is not possible to act on the causes. This characterization will allow the reflection on the consequences of the effect, which will allow to assess its severity using the defined criteria. It is intended to briefly describe the effect of the risk previously identified in order to better identify the critical impact on its activity.

Finally, the risk identified in (8), described in (9) and with the effect identified in (10), it is assessed in columns (11) of the PCMM.

Risk assessment is carried out using two criteria: Probability (P) and Severity (S). According to [2], the score criteria to be used in the estimation of Probability (P) and Severity (S) are proposed in **Tables 2** and **3**.

Then, for each risk or failure mode, its Probability (P) is given it a score. After, the analysis of the consequences and its effects, the same is done for Severity (S).

The scale of these scores should be assigned based on our experience with the activity in question.

The Risk Number (RN) classifies the assessed risk. Thus, if the Probability (P) and Severity (S) scores are multiplied, we obtain the RN in each case. Using **Table 4** found in [2], the RN that we can obtain vary between the minimum value 1 and the maximum value 16.

Therefore, high risk is classified when the NB is higher than 9, medium risk is classified when the NB is 9 and the low risk is classified when the NB is lesser than 9.

Probability		
Category	**Description**	**Score**
High	Occurs often.	4
Medium	Probably, it has occurred several times.	3
Low	Probably, it has already occurred.	2
Remote	Probably, but never occurred.	1

Table 2.
Risk probability criteria (P) [2].

Severity		
Category	**Description**	**Score**
High	Requires the re-inspection/rework the whole lot	4
Medium	Requires the re-inspection/rework part of the lot	3
Low	Requires adjustments in inspection/inspection/testing	2
Negligible	Does not require specific actions	1

Table 3.
Risk severity criteria (G) [2].

Risk Number		**Severity (S)**			
		1	2	3	4
Probability (P)	1	*1*	*2*	*3*	*4*
	2	*2*	*4*	*6*	*8*
	3	*3*	*6*	*9*	*12*
	4	*4*	*8*	*12*	*16*

Table 4.
Risk number [2].

With the classification of the RN done, the appreciation of events is began whose outputs are not the desired. Thus, the undesired outputs will be characterized, making it possible to make revision the inspections/tests identified and to adopt another type of control, measurement and monitoring.

In the column (12) the RN high, medium or low is identified and whether adequately is controlled or not.

However, according to the criteria defined in [2], the risk is adequately controlled (Yes) if the NR is lesser than 9 (for medium or low risks) and uncontrolled (No) if NR is higher than 9. For the uncontrolled risks (No) are must indicate the action required to mitigate them.

In the column (13) of the PCMM the actions to mitigate uncontrolled risk are indicated, making it adequately controlled.

3.2 Risk assessment with failure and effect modes analysis (FMEA)

In **Table 5** we see the template of a PCMM with FMEA that takes too the form of a matrix, constituted by a set of columns and rows containing information relating to the same items already explained above.

Using the FMEA technique for risk assessment, in the preparation of the PCMM we can use the following guidelines:

On the PCMM with FMEA form the fields (1) to (10) are the same as those of PCMM with LCM form. Field (11) is used to assess the risk identified in (8), described in (9) and with the effects identified in (10). Here, risk assessment will be carried out according to the three FMEA criteria, described below.
The classification of risk depends on the combination of probability (P), gravity (S) and detection (D).

Thus, for each risk or failure mode, we must analysis he consequences of their effects and to estimate its Severity (S). After to analysis its Probability (P) we must assign a it a score. Then, we must assign the Detection Index (D) to order the number priority of risk (NPR).

DOI: http://dx.doi.org/10.5772/intechopen.93957

Logo	PCMM - Plan Control, Measurement and Monitoring	Revision:	
		Page:	1/1
	Process/Work:		

Item (1)	Activities (1)	Inspection/ Testing (2)	Purposes (3)	Acceptance criteria (4)	Frequency/ Sampling (5)	Responsible (6)	Re cords (7)	Risks (8)	Description (9)	Effects (10)	Risk Assessment (11)				Controlled Risks? (12)	Actions (13)
											P	S	D	RPN		

Logo				PCMM - Plan Control, Measurement and Monitoring								Revision:	
												Page:	1/1
						Process/Work:							
Item (1)	**Activities (1)**	**Inspection/ Testing (2)**	**Purposes (3)**	**Acceptance criteria (4)**	**Frequency/ Sampling (5)**	**Responsible (6)**	**Re cords (7)**	**Risks (8)**	**Description (9)**	**Effects (10)**	**Risk Assessment (11)** P S D RPN	**Controlled Risks? (12)**	**Actions (13)**
Prepared by:		Date: _/_/_		Verified by:		Date: –/–/–			Responsible:			Date: –/–/–	

Table 5.
PCMM template with FMEA.

Therefore, it is necessary to establish the criteria to be used in the estimation of Probability (P), Severity (S) and Detection (D) values. **Table 2** shows the values of Probability (P), **Table 3** shows the values of Severity (S) and **Table 6** shows the values for Detection (D).

For the assignment of the scale of these scores, the experience we have of the activity under analysis is too important. But we are looking for is the risk priority order. Thus, if the Probability, Severity and Detection scores are multiplied, we will obtain the Risk Priority Number (RPN) in each case.

In accordance with **Table 4** we can classify High Risk as values higher than 9, Medium Risk values of 9 and Low Risk values lower than 9. In each class of risk, we must appreciate the ranking of the priority, namely, which are the most priority events where the outputs may not be desired.

Fields (12) and (13) on the PCMM with FMEA form are the same too as the fields explained above (PCMM with LCM). Thus, according to the criteria defined for the Risk Number, in field (12) we must consider the Risk appropriately controlled if the RN is ≤ 9 (for Medium or Low Risks), otherwise we should consider it uncontrolled (N). For the uncontrolled risks. Then, we must then indicate the action required to mitigate them. In field (13) we should indicate what actions are necessary to mitigate the most priority uncontrolled Risk, making it adequately controlled.

4. Application examples

Then, a brief presentation of a PCMM used in a works is made. The objective is to present two practical examples of the use of PCMM with risk assessment, the first under the LCM technique and the second under the FMEA technique, both examples from the methodology above mentioned.

4.1 PCMM with LCM

This case PCMM were prepared for the critical activities of the building works. However, **Table 7** shows an excerpt of the PCMM with LCM prepared for the execution of reinforced concrete beams of the building's structure.

The methodology set out above was followed in preparing of the PCMM with LCM, and risk assessment was made under Likelihood/Consequence Matrix.

The use of this procedure allows, on the one hand to plan the inspections and tests to make the critical activities, so as to control, measure and monitor all the work, on the other hand, it helps mitigate the risk in the events with a negative impact on the development of work.

It is observed that the planned control for the activity of the "Installation False-work" has the highest risk number (RN = 9), and it is classified as medium risk.

Detection		
Category	**Description**	**Score**
High	High difficulty detection	3
Moderate	Medium difficulty detection	2
Low	Low difficulty detection	1

Table 6.
Risk detection criteria (D).

Logo	PCMM - Plan Control, Measurement and Monitoring													Revision:
													Page:	1/1
	Process/Work: Execution of Beams													
Item (1)	Activities (1)	Inspection/ Testing (2)	Purposes (3)	Acceptance criteria (4)	Frequency/ Sampling (5)	Responsible (6)	Re cords (7)	Risks (8)	Description (9)	Effects (10)	Risk Assessment (11)		Controlled Risks? (12)	Actions (13)
											P	S	RN		
1	Receipt Falsework	Visual inspection	C with Purchase Order	Purchase Oder	By Delivery By Lot	Overseer	Control Sheet	NC with Purchase Order	Variation in Quantities; Deviations in type Falsework	Deviation in Activity 3	2	2	4	Yes	NA
2	Receipt Reinforcement	Visual and Metric inspection	C Steel Class and ϕ	Project; Steel specification for concrete reinforcement	By Delivery By Lot	Overseer	Control Sheet	NC with Specification	Deviation in Classes and ϕ of steel	Deviations in Activity 4	1	3	3	Yes	NA
3	Installation Falsework	Visual and Metric inspection	C of Verticality, Planimetry and Stability	Structural Design	By Beam	Overseer	Control Sheet	NC of Verticality, Planimetry and Stability	Instability of the beam	Deviations in activities 4 and 6	3	3	9	Yes	NA
4	Installation Reinforcement	Visual and Metric inspection	C of the mooring and positioning	Structural Design	By Beam	Overseer and Construction Engineer	Control Sheet	NC of mooring and positioning	Deformation and instability of the Beam	Deviations in Activity 6	2	3	6	Yes	NA
5	Receipt Concrete	Visual inspection and Consistency Testing	C with Purchase Order and Class Consistency	Purchase Order Structural Design	By delivery	Overseer and Construction Engineer	Control Sheet	NC of Purchase Order and NC Class Consistency	Deviations in Concrete Class Deviations in Consistency	Activities 7 stop	2	3	6	Yes	NA

Logo		PCMM - Plan Control, Measurement and Monitoring												Revision:	
														Page:	1/1
						Process/Work: Execution of Beams									
Item (1)	**Activities (1)**	**Inspection/ Testing (2)**	**Purposes (3)**	**Acceptance criteria (4)**	**Frequency/ Sampling (5)**	**Responsible (6)**	**Re cords (7)**	**Risks (8)**	**Description (9)**	**Effects (10)**	**Risk Assessment (11)**			**Controlled Risks? (12)**	**Actions (13)**
											P	**S**	**RN**		
6	Placement Concrete	Visual inspection	C of Concrete Placement	Structural Design	By Placement Concrete	Overseer and Construction Engineer	Control Sheet	NC of Concrete Placement	Deviations in the final quality of the Beam	Pathologies in the Beam	1	5	5	Yes	NA
...	...														
Prepared by:		Date:		Verified by:		Date:			Responsible:					Date:	
		_/_/_				–/–/–								–/–/–	

C - Compliance; NC - Not compliance; NA - Not Applicable.

Table 7.
PCMM with LCM for beams.

Logo			PCMM - Plan Control, Measurement and Monitoring												Revision:	
															Page:	1/1
						Process/Work: Execution of Beams										
Item (1)	Activities (1)	Inspection/ Testing (2)	Purposes (3)	Acceptance criteria (4)	Frequency/ Sampling (5)	Responsible (6)	Re cords (7)	Risks (8)	Description (9)	Effects (10)	Risk Assessment (11)				Controlled Risks? (12)	Actions (13)
											P	S	D	RPN		
1	Receipt Falsework	Visual inspection	C with Purchase Order	Purchase Oder	By Delivery By Lot	Overseer	Control Sheet	NC with Purchase Order	Variation in Quantities; Deviations in type Falsework	Deviation in Activity 3	2	2	1	4	Yes	NA
2	Receipt Reinforcement	Visual and Metric inspection	C Steel Class and ϕ	Project; Steel specification for concrete reinforcement	By Delivery By Lot	Overseer	Control Sheet	NC with Specification	Deviation in Classes and ϕ of steel	Deviations in Activity 4	1	3	1	3	Yes	NA
3	Installation Falsework	Visual and Metric inspection	C of Verticality, Planimetry and Stability	Structural Design	By Beam	Overseer	Control Sheet	NC of Verticality, Planimetry and Stability	Instability of the beam	Deviations in activities 4 and 6	3	3	3	27	Yes	NA
4	Installation Reinforcement	Visual and Metric inspection	C of the mooring and positioning	Structural Design	By Beam	Overseer and Construction Engineer	Control Sheet	NC of mooring and positioning	Deformation and instability of the Beam	Deviations in Activity 6	2	3	3	18	Yes	NA
5	Receipt Concrete	Visual inspection and Consistency Testing	C with Purchase Order and Class Consistency	Purchase Order Structural Design	By delivery	Overseer and Construction Engineer	Control Sheet	NC of Purchase Order and NC Class Consistency	Deviations in Concrete Class Deviations in Consistency	Activities 7 stop	2	3	3	18	Yes	NA

DOI: http://dx.doi.org/10.5772/intechopen.93957

Logo				PCMM - Plan Control, Measurement and Monitoring											Revision:	
															Page:	1/1
							Process/Work: Execution of Beams									
Item (1)	**Activities (1)**	**Inspection/ Testing (2)**	**Purposes (3)**	**Acceptance criteria (4)**	**Frequency/ Sampling (5)**	**Responsible (6)**	**Re cords (7)**	**Risks (8)**	**Description (9)**	**Effects (10)**	**Risk Assessment (11)**				**Controlled Risks? (12)**	**Actions (13)**
											P	**S**	**D**	**RPN**		
6	Placement Concrete	Visual inspection	C of Concrete Placement	Structural Design	By Placement Concrete	Overseer and Construction Engineer	Control Sheet	NC of Concrete Placement	Deviations in the final quality of the Beam	Pathologies in the Beam	1	5	3	15	Yes	NA
...	...															
Prepared by:		Date: _/_/_		Verified by:		Date: –/–/–				Responsible:					Date: –/–/–	

Table 8.
PCMM with FMEA for beams.

Given the set decision rule in the methodology, this risk is controlled and does not require mitigation action.

It is seen too that the planned control for the activity "Receipt Reinforcement" has the lowest risks number (RN = 3), and it is classified as low risk. With the set decision rule defined in the methodology, this risk is also controlled and does not require any mitigation action.

It is seen that the planned control for the activity of the "Installation Falsework" has the highest risk number (RN = 9), and it is classified as medium risk. Given the set decision rule in the methodology, this risk is controlled and does not require mitigation action.

It is observed that the planned control for the activity "Receipt Reinforcement" has the lowest risks number (RN = 3), and it is classified as low risk. With the set decision rule defined in the methodology, this risk is also controlled and does not require any mitigation action.

4.2 PCMM with FMEA

Keeping the form of the PCMM with LCM shown above, the **Table 8** shows the part of the PCMM with FMEA prepared for the execution of reinforced concrete beams of the building's structure.

The methodology set out above was followed in preparing PCMM with FMEA and risk assessment was made under Failure Mode and Effects Analysis (FMEA).

The use of this procedure also allows, on the one hand to plan the inspections and tests to make the critical activities, so as to control, measure and monitor all the work, on the other hand, it helps prevent the risk in the occurrence of monitoring events with a negative impact on the development of work.

It is seen that the planned control for the activity of the "Installation Falsework" continues to have the highest risk number and the highest risk priority number (RPN = 27). Since it is classified as medium risk, according to the set decision rule defined in the methodology, this risk is controlled and does not require any mitigation action.

It is observed that the planned control for the activity "Receipt Reinforcement" has the lowest risk number and the lowest risk priority number (RPN = 3), it being classified as low risk. With the set decision rule defined in the methodology, this risk is also controlled and does not require any mitigation action.

5. Conclusion

It was concluded the methodology presented for the preparation of Control, Measurement and Monitoring Plans (PCMM), can follow the approach on risk-based thinking and can help to assurance the compliance with technical and regulatory of the works. It was concluded too that measurement and monitoring process can promote the risks and failures assessment and can demonstrate the conformity of the works and can too improve operational efficiency.

The PCMM with risk assessment helps to identification the need for risk mitigation actions in the control of the works. Therefore, the use of risk assessment techniques is important in the planning and control of the works processes.

It was observed that Likelihood/Consequence Matrix (LCM) and Failure Mode and Effects Analysis (FMEA) are two techniques of the risk assessment applicable to PCMM, and introducing the probability, consequences and detection indices, they allow risk classification and priority of their mitigation.

It also was concluded that the use of the proposed methodology for of the PCMM can be a solution to prevent defects and reworks, since it allows you to easily identify which trials and inspections will be implement on the control, measurement and monitoring critical activities of the works.

If we are used to risk assessment for identify from the sources, events and causes and its possible consequences, then we can determine the level of risk and its acceptance or tolerance. Finally, the risk assessment applied to PCMM helps to indicate which actions are necessary to mitigate uncontrolled risks.

With the examples presented, it was possible to conclude that defects and rework of the works do not have to be accepted as inevitable or even as certainties and should be considered as a permanent challenge to the management of work.

Author details

M. Rosário Oliveira†
ISEP - School of Engineering Polytechnic of Porto, Porto, Portugal

*Address all correspondence to: mro@isep.ipp.pt

† ORCID 0000-0001-8149-7351

References

[1] Romão T.G. Evolução do Sector da Construção em Portugal - Aplicação do Modelo Structure-Conduct-Performance [Dissertação]. Instituto Superior Técnico-IST. Lisboa, 2015.

[2] Oliveira M.R. Plans of Control, Measurement and Monitoring with Risk Assessment Application to Rehabilitation works. Structural Integrity Procedia. vol. 5, pp. 1129-1135, 2017. DOI: 10.1016/j.prostr.2017.07.016.

[3] Ali M.C. Exploring the Potential of Integration Quality Assessment System in Construction (QLASSIC) With ISO 9001 Quality Management System (QMS). International Journal for Quality Research, vol. 8, nº 1, pp. 73-86, 2014. ISSN 1800-6450.

[4] Oliveira M.R. A Gestão da Qualidade na Construção e a Gestão do Risco. ISEP Moodle, Porto, 2016.

[5] NP EN ISO 9001:2015 - Sistemas de Gestão da Qualidade - Requisitos (ISO 9001:2015). Instituto Português da Qualidade (IPQ), Caparica, 2015.

[6] DNP ISO Guia 73:2011 - Gestão do risco - Vocabulário (ISO Guide 73: 2009). Instituto Português da Qualidade (IPQ), Caparica, 2011.

[7] NP EN 31010:2016 - Gestão do risco - Técnicas de apreciação do risco (ISO/ IEC 31010:2009). Instituto Português da Qualidade (IPQ), Caparica, 2016.

[8] Nuchpho P, Nansaarng S, Pongpullponsak A. Risk Assessment in the Organization by Using FMEA Innovation: A Literature Review. King Mongkut's University Technology Thonburi, Thailand, 2014.

[9] M. Abdelgawad M, Fayek A.R. Risk Management in the Construction Industry Using Combined Fuzzi FMEA and Fuzzy AHP. Journal of Construction Engineering and Management, vol. 136, nº 9, pp. 1028-1036, 2010. DOI: 10.1061/(ASCE)CO.1943-7862. 0000210.

[10] Wang H, Deng X, Zhang Z, Jiang W. A New Failure Mode and Effects Analysis Method. IEEE Acess ISSN 2169-3536, vol. 7, p. 12, 2019. DOI: 10.1109/ACCESS.2019.2923064.

[11] Silva S.R.C, Fonseca M, Brito J. Metodologia FMEA e sua Aplicação à Construção de Edificios. LNEC, QIC2006, Lisboa, 2006.

[12] Huang J, You J-X, Liu H-C, Song M-S. Failure mode and effect analysis improvement: A systematic literature. Reliability Engineering and System Safety ISSN 0951-8320, vol. 199, p. 12, 2020. DOI: 10.1016/j. ress.2020.106885.

[13] Chen J. K. Utility Priority Number Evaluation for FMEA. Journal of Failure Analysis and Prevention, vol. 7, pp. 321-328, 2007. DOI: 10.1007/ s11668-007-9060-2.

[14] Oliveira M.R. Planos de Controlo, Medição e Monitorização – Procedimento. ISEP Moodle, Porto, 2018.

Section 2

Bridge Engineering

Chapter 5

Analytic and Numerical Results of Bending Deflection of Rectangular Composite Plate

Louay S. Yousuf

Abstract

In this chapter, the derivation of analytic formulation of bending deflection has been done using the theory of classical laminate plate. The method of Navier and Levy solutions are used in the calculation. The composite laminate plate is exposed to out-off plane temperatures and combined loading. The temperature gradient of thermal shock is varied between $60C^{\circ}$ and $-15C^{\circ}$. The combined loading are the bending moment (M_o) in the y-direction and in-plane force (N_{xx}) in the x-direction. The in-plane force (N_{xx}) has a great effect on the bending deflection value within a (95.842%), but the bending moment (M_o) has a small effect on the bending deflection value in the rate of (4.101%). The results are compared and verified for central normal deflection.

Keywords: classical plate theory, composite laminate plate, temperature affect, combined loading

1. Introduction

The effect of temperature and combined loading on composite plate is one of the primary life limiting factors of a bridge engineering application. This chapter will consider the structural evaluation of the localized effect in the bridge engineering. The application of bridge engineering can be found in a structural bridge deck panel. Ray studied the fiber-matrix debonding by applying the thermal shock of thermal fatigue, taking into account the conditioning time. He performed a three-point bending test on glass fiber reinforced with unsaturated polyester and epoxy resin composites in which it exposed to 75 C° of the temperature gradient [1]. Hussein and Alasadi used a numerical analysis of stress and strain values of angle-ply with four-layered symmetric laminated plate under the effect of force resultant (N_{xx}) and bending moment (M_{xx}) graphically. He predicted the material properties of the multilayered plate of the reinforcement fibers of E-glass and epoxy resin [2]. Yousuf et al. evaluated the dynamic analysis of normal deflection, taking into consideration the effect of thermal fatigue beside the effect of bending moment (M_o) and in-plane force (N_{xx}). The composite laminate plate of woven roving fiber glass and polyester were exposed to 75 C° of the temperature gradient. A composite laminate plate with fiber volume fraction $(v_f = 25.076\ \%)$ was selected [3]. Wang et al. applied the thermal cycles in the temperature range between (80 C° and -40 C°) on different plys of glass fiber/epoxy matrix composites. Scanning electron microscopy (SEM) images showed that after 180 of thermal cycles, the bonding effect of glass fiber and epoxy matrix became worse, leading to the decrease in mechanical properties [4].

IntechOpen

Khashaba et al. investigated the mechanical properties of $[0]_8$ of woven glass fiber reinforced polyester (GFRP) composites under monotonic and combined tension/bending loading [5, 6]. Yousuf reduced the vibration properties of composite material under the variation of combined temperatures (60 C^o to -15 C^o) using three types of boundary conditions. The free vibration test was carried out for (5, 10, 15, 20, 25, and 30) minutes [7]. Moufari proposed several numerical simulations to describe the interaction between thermal and mechanical stresses. The estimation damage modes of carbon/epoxy laminate plate has been achieved due to thermal cyclic loading. Zhen and Xiaohui proposed an analytic model of Reddy-type higher-order plate theory for simply supported plates based on thermal and mechanical combined loading [8]. In this work, the analytic derivation of bending deflection has been done by using the theory of classical laminate plate. Levy and Navier solutions are used to describe the theory of bending deflection by taking into consideration the use of simply supported boundary condition from all edges. In our point of view, the analytic derivation of normal deflection under the effect of temperature and combined loading has not been studied.

2. Equations of motion in terms of displacements

The stress and strain relationship is varied through the laminate thickness, as indicated in Eq. (1):

$$\begin{bmatrix} \sigma_{xx} \\ \sigma_{yy} \\ \sigma_{xy} \end{bmatrix} = \begin{bmatrix} Q_{11} & Q_{12} & 0 \\ Q_{12} & Q_{22} & 0 \\ 0 & 0 & Q_{66} \end{bmatrix} * \left[\begin{bmatrix} \varepsilon_{xx} \\ \varepsilon_{yy} \\ \gamma_{xy} \end{bmatrix} - \begin{bmatrix} \alpha_1 \\ \alpha_2 \\ 0 \end{bmatrix} * \Delta T \right] \tag{1}$$

The general bending equation of rectangular plate is as below:

$$\frac{\partial^2 M_{xx}}{\partial x^2} + 2\frac{\partial^2 M_{xy}}{\partial x \partial y} + \frac{\partial^2 M_{yy}}{\partial y^2} = 0 \tag{2}$$

By taking into account the temperature effect, the mechanical and thermal bending moments are:

$$\begin{bmatrix} M_{xx} \\ M_{yy} \\ M_{xy} \end{bmatrix} = \begin{bmatrix} M_{xx}^{Mech.} \\ M_{yy}^{Mech.} \\ M_{xy}^{Mech.} \end{bmatrix} - \begin{bmatrix} M_{xx}^{Ther.} \\ M_{yy}^{Ther.} \\ M_{xy}^{Ther.} \end{bmatrix} \tag{3}$$

where

$$\begin{bmatrix} M_{xx}^{Mech.} \\ M_{yy}^{Mech.} \\ M_{xy}^{Mech.} \end{bmatrix} = \begin{bmatrix} B_{11} & B_{12} & 0 \\ B_{12} & B_{22} & 0 \\ 0 & 0 & B_{66} \end{bmatrix} * \begin{bmatrix} \frac{\partial u_o}{\partial x} \\ \frac{\partial v_o}{\partial y} \\ \frac{\partial u_o}{\partial y} + \frac{\partial v_o}{\partial x} \end{bmatrix} - \begin{bmatrix} D_{11} & D_{12} & 0 \\ D_{12} & D_{22} & 0 \\ 0 & 0 & D_{66} \end{bmatrix} * \begin{bmatrix} \frac{\partial^2 w_o}{\partial x^2} \\ \frac{\partial^2 w_o}{\partial y^2} \\ 2\frac{\partial^2 w_o}{\partial x \partial y} \end{bmatrix} \tag{4}$$

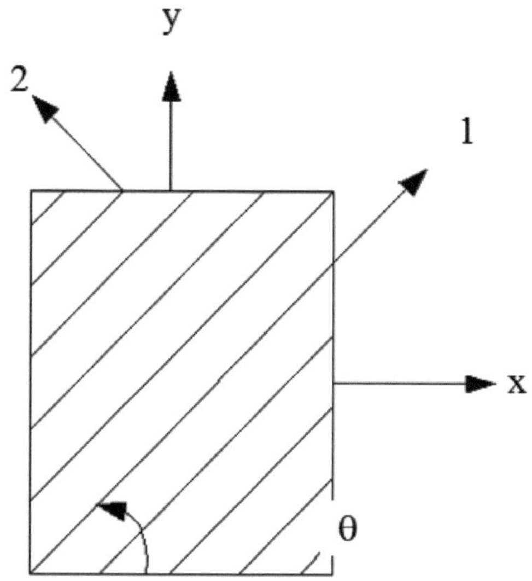

Figure 1.
Lamina of arbitrary of principal material direction.

And,

$$\begin{bmatrix} M_{xx}^{Ther.} \\ M_{yy}^{Ther.} \end{bmatrix} = \sum_{k=1}^{N} \begin{bmatrix} Q_{11} & Q_{12} \\ Q_{12} & Q_{22} \end{bmatrix}^{k} * \begin{bmatrix} \alpha_1 \\ \alpha_2 \end{bmatrix}^{k} * \int_{z_k}^{z_{k+1}} \Delta T z dz \tag{5}$$

It can be assumed that all layers have $(\Theta = 0)$, and the same thickness $(B_{ij} = 0)$, as indicated in **Figure 1**. Substitute Eq. (4) and into Eq. (3), it can be obtained:

$$D_{11}\left(\frac{\partial^4 w_o}{\partial x^4}\right) + 2(D_{12} + 2D_{66})\left(\frac{\partial^4 w_o}{\partial x^2 \partial y^2}\right) + D_{22}\left(\frac{\partial^4 w_o}{\partial y^4}\right) + N_{xx}\left(\frac{\partial^2 w_o}{\partial x^2}\right) = -\left(\frac{\partial^2 M_{xx}^{Ther.}}{\partial x^2} + \frac{\partial^2 M_{yy}^{Ther.}}{\partial y^2}\right) \tag{6}$$

3. Formulation of bending deflection distribution using Navier solution

The normal deflection distribution is derived based on the solution of classical laminate plate theory using Navier equation. Navier solution assumed that the boundary condition is simply supported from all edges under the effect of temperature (ΔT) and in-plane force (N_{xx}). It can be assumed that the temperature is varied linearly through the plate thickness, as in below:

$$T_{mn}(z) = \frac{4}{ab}\int_0^a\int_0^b \Delta T(x,y,z)\, sin\,(\alpha_m x)\, sin\,(\beta_n y) dxdy \tag{7}$$

where.

T_1 is the out-off plane uniform temperature when the heat source is applied through the plate thickness.

And,

$$\alpha_m = \frac{m\pi}{a}, m = 1, 2, 3, \dots, \infty \tag{8a}$$

$$\beta_n = \frac{n\pi}{b}, n = 1, 2, 3, \dots, \infty \tag{8b}$$

By integrating Eq. (7) with respect to (x) and (y), the temperature distribution through the plate thickness is:

$$T_{mn}(z) = \frac{16T_1 z}{mn\pi^2} \tag{9}$$

The thermal bending moment is defined as in the following:

$$\begin{bmatrix} M_{xx}^{Ther.} \\ M_{yy}^{Ther.} \end{bmatrix} = \sum_{m=1}^{\infty}\sum_{n=1}^{\infty} \begin{bmatrix} M_{mn}^{(1)} \\ M_{mn}^{(2)} \end{bmatrix} sin\,(\alpha_m x)\, sin\,(\beta_n y) \tag{10}$$

where

$$M_{mn}^{(1)} = \sum_{k=1}^{N} \int_{z_k}^{z_{k+1}} (Q_{11}\alpha_1 + Q_{12}\alpha_2) T_{mn}(z) z dz \tag{11a}$$

$$M_{mn}^{(2)} = \sum_{k=1}^{N} \int_{z_k}^{z_{k+1}} (Q_{12}\alpha_1 + Q_{22}\alpha_2) T_{mn}(z) z dz \tag{11b}$$

The general solution of normal deflection for simply supported boundary condition from all edges is:

$$w_o = \sum_{m=1}^{\infty}\sum_{n=1}^{\infty} w_{mn}\, sin\,(\alpha_m x)\, sin\,(\beta_n y) \tag{12}$$

Substitute Eq. (12) and Eq. (10) into Eq. (6), the solution of bending deflection is illustrated in the following equation:

$$w_o(x, y) = \frac{16T_1}{3\pi^2} \sum_{m=1}^{\infty}\sum_{n=1}^{\infty} \frac{\left(A_1\alpha_m^2 + A_2\beta_n^2\right) sin\,(\alpha_m x)\, sin\,(\beta_n y)}{mn\left(D_{11}\alpha_m^4 + 2(D_{12} + 2D_{66})\alpha_m^2\beta_n^2 + D_{22}\beta_n^4 - N_{xx}\alpha_m^2\right)} \tag{13}$$

where

$$A_1 = \sum_{k=1}^{N} (Q_{11}\alpha_1 + Q_{12}\alpha_2)\left(z_{k+1}^3 - z_k^3\right) \tag{14a}$$

$$A_2 = \sum_{k=1}^{N} (Q_{12}\alpha_1 + Q_{22}\alpha_2)\left(z_{k+1}^3 - z_k^3\right) \tag{14b}$$

4. Formulation of bending deflection distribution using levy solution

The theory of classical laminate plate of Levy solution is used to derive the solution of normal deflection. The Levy solution assumed that the variation of the bending deflection should be along the x-axis. Levy solution can be used on any type of boundary condition which gives flexibility on any type of loading such as (ΔT), in-plane force (N_{xx}), and bending moment (M_o). As mentioned in the previous section that the temperature is varied linearly through the plate thickness, as below:

$$T_m(z) = \frac{2}{a}\int_0^a \Delta T(x,z)\,sin\,(\alpha_m x)dx \tag{15}$$

where

$$\alpha_m = \frac{m\pi}{a}, m = 1, 2, 3, \dots, \infty \tag{16}$$

By integrating Eq. (15) with respect to (x), the temperature distribution through the plate thickness is:

$$T_m(z) = \frac{4T_1 z}{m\pi} \tag{17}$$

Ignore the variation of thermal bending moment and normal deflection along y-axis, Eq. (6) will be:

$$D_{11}\left(\frac{\partial^4 w_o}{\partial x^4}\right) + N_{xx}\left(\frac{\partial^2 w_o}{\partial x^2}\right) = -\frac{\partial^2 M_{xx}^{Ther.}}{\partial x^2} \tag{18}$$

As mentioned earlier, the thermal bending moment is varied along x-axis, as below:

$$\left[M_{xx}^{Ther.}\right] = \sum_{m=1}^{\infty}\left[M_m^{(1)}\right] sin\,(\alpha_m x) \tag{19}$$

where

$$M_m^{(1)} = \sum_{k=1}^{N}\int_{z_k}^{z_{k+1}} (Q_{11}\alpha_1)T_m(z)z dz \tag{20}$$

The solution of normal bending deflection is as below:

$$w_o(x,y) = w_o(x,y)_H + w_o(x)_P \tag{21}$$

To find $w_o(x)_P$:

$$w_o(x,t)_p = \sum_{m=1}^{\infty} w_m\,sin\,(\alpha_m x) \tag{22}$$

Substitute Eq. (19) and Eq. (22) into Eq. (18) to obtain the particular solution of bending deflection along x-axis, $w_o(x)_P$:

$$w_o(x)_p = \frac{4T_1 A_1}{3\pi}\sum_{m=1}^{\infty}\frac{\alpha_m^2\,sin\,(\alpha_m x)}{m\left(D_{11}\alpha_m^4 - N_{xx}\alpha_m^2\right)} \tag{23}$$

where

$$A_1 = \sum_{k=1}^{N} (Q_{11}\alpha_1)\left(z_{k+1}^3 - z_k^3\right) \tag{24}$$

To find $w_o(x,y)_H$, Eq. (6) will be:

$$D_{11}\frac{\partial^4 w_o}{\partial x^4} + 2(D_{12} + 2D_{66})\frac{\partial^4 w_o}{\partial x^2 \partial y^2} + D_{22}\frac{\partial^4 w_o}{\partial y^4} + N_{xx}\frac{\partial^2 w_o}{\partial x^2} = 0 \tag{25}$$

The solution of Eq. (25) is as below:

$$w_o(x,y)_H = \sum_{m=1}^{\infty} Y_m \, sin\,(\alpha_m x) \tag{26}$$

Substitute Eq. (26) into Eq. (25), to obtain the homogeneous solution of Eq. (25) along x- and y-directions:

$$w_o(x,y)_H = \sum_{m=1}^{\infty} [N_1 \, cosh\,(\alpha y) + N_2 \, sinh\,(\alpha y) + N_3 \, cos\,(\beta y) + N_4 \, sin\,(\beta y)] \; sin\,(\alpha_m x) \tag{27}$$

Substitute Eq. (27) and Eq. (23) into Eq. (21) to obtain the general solution of normal bending deflection, as indicated below:

$$w_o(x,y) = \sum_{m=1}^{\infty} [N_1 \, cosh\,(\alpha y) + N_2 \, sinh\,(\alpha y) + N_3 \, cos\,(\beta y) + N_4 \, sin\,(\beta y)] \; sin\,(\alpha_m x) + \frac{4T_1 A_1}{3\pi} \sum_{m=1}^{\infty} \frac{\alpha_m^2 \, sin\,(\alpha_m x)}{m\left[D_{11}\alpha_m^4 - N_{xx}\alpha_m^2\right]} \tag{28}$$

The simply supported boundary conditions from all edges are assumed and the constants ($N_1, N_2, N_3, and\ N_4$) are as below:

$$N_1 = -\frac{\beta^2 H}{(\alpha^2 + \beta^2)} - \frac{4M_o}{(\alpha^2 + \beta^2) m\pi D_{22}} \tag{29}$$

$$N_2 = \frac{cosh\,(\alpha b)\beta^2 H}{(\alpha^2 + \beta^2)\, sinh\,(\alpha b)} + \frac{4\, cosh\,(\alpha b) M_o}{(\alpha^2 + \beta^2)\, sinh\,(\alpha b) m\pi D_{22}} - \frac{(D_{12}\alpha_m^2 + \beta^2 D_{22})H}{(\alpha^2 + \beta^2)\, sinh\,(\alpha b) D_{22}} - \frac{\left(\frac{4M_o}{m\pi} - D_{12}\alpha_m^2 H\right)}{(\alpha^2 + \beta^2)\, sinh\,(\alpha b) D_{22}} \tag{30}$$

$$N_3 = -\frac{\alpha^2 H}{(\alpha^2 + \beta^2)} + \frac{4M_o}{(\alpha^2 + \beta^2) m\pi D_{22}} \tag{31}$$

$$N_4 = \frac{cos\,(\beta b)\alpha^2 H}{(\alpha^2 + \beta^2)\, sin\,(\beta b)} - \frac{4\, cos\,(\beta b) M_o}{(\alpha^2 + \beta^2)\, sin\,(\beta b) m\pi D_{22}} - \frac{(\alpha^2 D_{22} - D_{12}\alpha_m^2)H}{(\alpha^2 + \beta^2) D_{22}\, sin\,(\beta b)} + \frac{\left(\frac{4M_o}{m\pi} - D_{12}\alpha_m^2 H\right)}{(\alpha^2 + \beta^2)\, sin\,(\beta b) D_{22}} \tag{32}$$

where

$$H = \frac{4T_1A_1\alpha_m^2}{3\pi m\left(D_{11}\alpha_m^4 - N_{xx}\alpha_m^2\right)} \tag{33}$$

5. Numerical simulation procedure

In this chapter, the finite element discretization is carried out by using ANSYS Ver. 18.2. (SHELL 132) element is used to mesh the composite laminate plate. SHELL 132 is defined by eight nodes having six degrees of freedom at each node to calculate the central normal deflection. In the simulation analysis, the central point of laminate plate is used to calculate the normal deflection. Always the convergence test is needed to determine the size of elements in which the value of normal bending deflection settles down. Finite element analysis of convergence curve defines the relationship between the grid interval and the analysis accuracy. Four types of combined loading is used such as: (temperature affect only (ΔT)), (temperature affect (ΔT) + Mo), (temperature affect (ΔT) + Nxx), and (temperature affect (ΔT) + Mo + Nxx). Multiple values of fiber volume fraction is used such as (25, 40, 50, 60, 70, and 80)%. **Table 1** shows the mechanical and thermal properties of the simulated materials.

ν_f	25.07%	40%	50%	60%	70%	80%
$E_x^c(GPa.)$	19.933	30.4038	37.41988	44.435	51.452	58.468
$E_y^c(GPa.)$	19.933	30.4038	37.41988	44.435	51.452	58.468
$E_z^c(GPa.)$	3.0896	3.81746	4.53322	5.5793	7.25302	10.3612
ν_{12}	0.3835	0.35098	0.32915	0.30732	0.2855	0.26366
$G_{12}(GPa.)$	1.07675	1.33379	1.5878	1.9614	2.5648	3.70468
$\rho_c(kg/m^3)$	1464.18	1686.48	1835.4	1984.32	2133.24	2282.16
$\alpha_x^c(1/C^\circ)$	25.746 E-6	21.6044 E-6	18.5098 E-6	15.3005 E-6	12.0234 E-6	8.70307 E-6
$\alpha_y^c(1/C^\circ)$	25.746 E-6	21.6044 E-6	18.5098 E-6	15.3005 E-6	12.0234 E-6	8.70307 E-6
$\alpha_z^c(1/C^\circ)$	10.5844 E-6	7.932 E-6	6.9852 E-6	6.3374 E-6	5.8663 E-6	5.5082 E-6
$k_x^c(W/mC^\circ)$	0.4533	0.622	0.735	0.848	0.961	1.074
$k_y^c(W/mC^\circ)$	0.4533	0.622	0.735	0.848	0.961	1.074
$k_z^c(W/mC^\circ)$	0.2174	0.2626	0.30068	0.3553	0.43418	0.55808
$C_p^c(J/kgC^\circ)$	768.139	780.8944	787.7133	793.5087	798.495	802.8304

Table 1.
Mechanical and thermal properties of the simulated materials.

6. Results and discussions

Figures 2 and **3** show the verification test of normal bending deflection using Levy and Navier solutions, taking into consideration ANSYS 18.2 results. The normal bending deflection decreased with the increasing of plate aspect ratio because of the increasing in plate bending stiffness under the temperature effect $(60C^\circ)$ and $(-15C^\circ)$ for fiber volume fraction (25.076%). The bending deflection value when

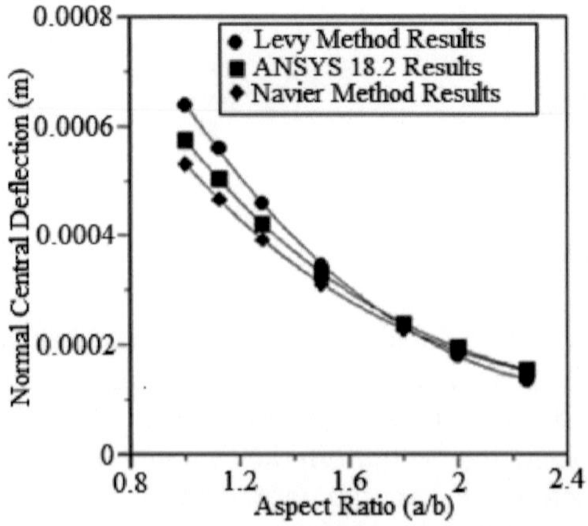

Figure 2.
Normal bending deflection varying with laminate plate aspect ratio under temperature effect (60C°).

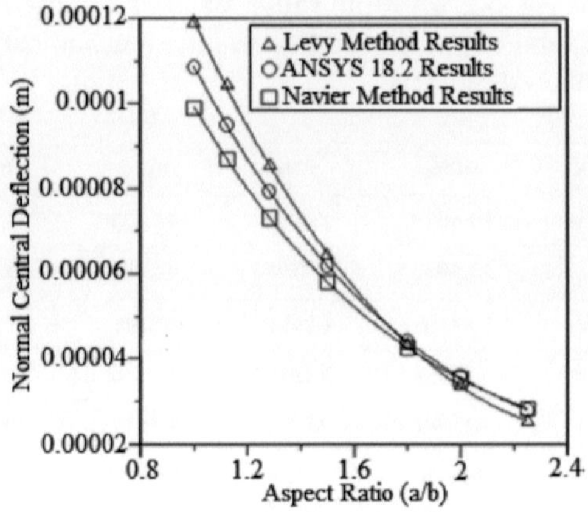

Figure 3.
Normal bending deflection varying with laminate plate aspect ratio under temperature effect (−15C°).

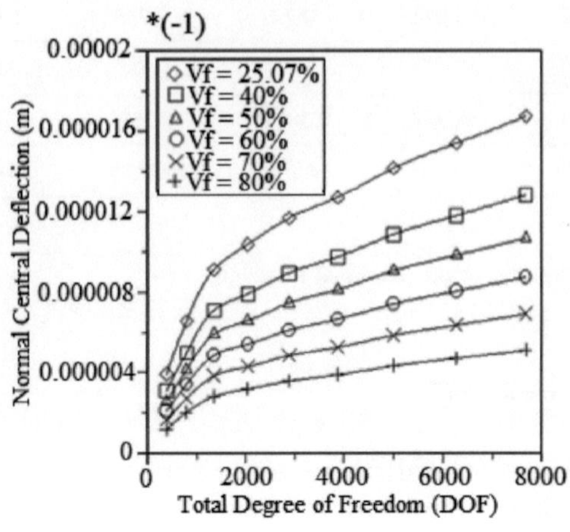

Figure 4.
Convergence test of normal deflection and total degrees of freedom under the effect of temperature (ΔT), *bending moment* (M_o), *and in-plane force* (N_{xx}).

Deflection	Levy method results	ANSYS 18.2 results	Percentage error (%)
(ΔT)	0.1853e-3	0.1880e-3	1.748
(ΔT)+ (M_o)	−0.1777e-3	−0.1882e-3	5.536
(ΔT)+ (N_{xx})	0.7704e-5	0.7108 e-5	7.736
(ΔT)+ (M_o)+(N_{xx})	−0.9859e-5	−0.9365e-5	5.010

Table 2.
Analytic and simulation verification of bending deflection under combined loading.

$(T = 60C^{\circ})$ is higher than the value of bending deflection when $(T = -15C^{\circ})$ because of the expansion and contraction through the plate laminate thickness.

Figure 4 shows the convergence test of normal bending deflection with total degrees of freedom for different fiber volume fractions using ANSYS software. The normal central deflection decrease with the increasing of fiber volume fraction under the effect of temperature (ΔT), bending moment (M_o), and in-plane force (N_{xx}).

Table 2 shows the analytic and simulation verification results of bending deflection under combined loadings for fiber volume fraction $\upsilon_f = 25.076\%$ and plate aspect ratio (1.8). The value of central deflection of the system with combined loading (ΔT) is higher than the others of combined loading. The deflection of system with combined loading $(\Delta T + M_o + N_{xx})$ and the system with loading $(\Delta T + N_{xx})$ is almost the same and in the opposite direction because bending moment has a small effect.

7. Conclusions

As mentioned in Introduction section, Levy and Navier solutions are used to describe the theory of bending deflection by taking into consideration the use of simply supported boundary condition from all edges. ANSYS software is used in the convergence test. The bending deflection value when $(T = 60C^{\circ})$ is higher than the value of bending deflection when $(T = -15C^{\circ})$ because of the expansion and contraction through the plate laminate thickness. The in-plane force (N_{xx}) has a great effect on the bending deflection value of composite laminate plate, but the bending moment (M_o) has a small effect on the bending deflection value. The normal deflection is decreased with the increasing of fiber volume fraction from 25.07% to 80% under the effect of (ΔT) and combined loading (M_o) + (N_{xx}). Moreover, the normal bending deflection is decreased with the increasing of aspect ratio from 0.8 to 2.4 under the effect of $(T = 60C^{\circ})$ and $(T = -15C^{\circ})$, respectively.

Nomenclature

α_1, α_2	thermal expansion coefficient in longitudinal and lateral directions, $1/C^{\circ}$.
ΔT	gradient uniform temperature, C°.
A_1, A_2	bending moment due to temperature, N.m/C°.
M_{xx}, M_{yy}, and M_{xy}	bending and twist moments, N.m.
Q_{ij}	reduced stiffness elements, N/m^2.
w_0	midplane deflection along z-direction.
z_k, z_{k+1}	upper and lower lamina surface coordinates along z-direction, m.

a, b	length of large and small spans of rectangular plate (m).
m, n	double trigonometric of Furrier series.
N	total number of layers.

Author details

Louay S. Yousuf
Department of Mechanical Engineering, San Diego State University, San Diego, CA, USA

*Address all correspondence to: louaysabah79@yahoo.com

References

[1] Ray B. Thermal shock and thermal fatigue on delamination of glass-fiber-reinforced polymeric composites. Journal of Reinforced Plastics and Composites. 2005;**24**(1):111-116

[2] Hussein EQ, Alasadi SJ. Experimental and theoretical stress analysis for composite plate under combined load. Journal of University of Babylon. Babylon University; 2018;**26**(1):129-140

[3] Yousuf LS, Jameel AN, AL-Sahib NKA. Verification of laminate composite plate simulation under combined loadings thermal stresses. Journal of Engineering. 2010;**16**(4): 6123-6142

[4] Wang D, Zhou X, Ge H, Liu Z, Liu H, Sun K. The influence of thermal fatigue on the properties of glass fiber/epoxy composites. Polymers & Polymer Composites. 2012;**20**(1-2):129-132

[5] Khashaba U, Aldousari S, Najjar I. Behavior of $[0]_8$ woven composites under combined bending and tension loading: Part-i experimental and analytical. Journal of Composite Materials. 2012;**46**(11):1345-1355

[6] Onyechi PC, Asiegbu KO, Chinenye AL. Effect of volume fraction on the mechanical properties of Periwinkle shell reinforced polyester composite (PRPC). American Journal of Mechanical Engineering and Automation. Open Science Publisher; 2015;**2**(1):1-15

[7] Yousuf LS. Time prediction of dynamic behavior of glass fiber reinforced polyester composites subjected to fluctuating varied temperatures. Al-Khwarizmi Engineering Journal. 2018;**5**(3):28-37

[8] Zhen W, Xiaohui R. Thermomechanical analysis of multilayered plates in terms of Reddy-type higher-order theory. Journal of Advanced Materials and Structures. Taylor & Francis Publisher; 2017;**24**(14): 1196-1205

Chapter 6

Temporary Works in Construction of Bridges Near Third Party Assets

Ganga Kasi V. Prakhya

Abstract

This paper gives a summary of the temporary works and methods applied to enable the construction of bridges near third party assets. The temporary structures have a significant impact on the cost, construction method and construction safety of the supported permanent structures. In the literature, there are many examples of how the temporary works could fail catastrophically and could endanger public life if hazard identification, risk assessment and quality checks are not carried out by competent people or organisations. A brief literature survey of the construction techniques is outlined here. This paper looks at more recent failures and draws some more lessons with a re-emphasis on the use of industry established processes, guidelines for preventing catastrophic events. The paper also describes case histories, where mitigation measures are implemented in order to ensure safety by means of independent checks, monitoring and back analysis.

Keywords: temporary works, monitoring, safety in construction, method statements, risk assessment, back analysis

1. Introduction

Most forms of bridge construction, whether preassembled or cast in-situ will require temporary works on site. These may include, depending on the type of bridge, temporary supports for precast girders or beams, box structures, and temporary staging for cast in-situ construction of the deck, and may also involve specialist operations for complex forms of bridge, e.g., post-tensioning. Many projects focus on optimising the concrete volumes and steel tonnage but a very few focus on how best to integrate temporary works and buildability concepts into the permanent design at the early stages.

Temporary works (TW) are the parts of a construction project that are needed to enable the permanent works to be built. Usually the TW are removed after use—e.g. access scaffolds, props, shoring, excavation support, falsework, and formwork, etc. It may be possible that sometimes the TW an integral part of the permanent works—e.g. props for excavation can be used as part of permanent steelworks, haul road foundations and crane or piling platforms may be used for hard standing or road foundations. In view of safety, it is very important that the same degree of care and attention is given to the design and construction of temporary works (TW) as to the design and construction of the permanent works. As TW may be in place for only a short during the construction phase of the project, there is a tendency to assume they are less important. Lack of care with design, selection, assembly, etc. leaves TW liable to fail or collapse [1]. This places people at risk of injury and can cause the project to be

IntechOpen

delayed. Therefore it is important to ensure that the methods, materials, and sequence of construction is thought through with the construction team during early stages and that the risks to the structure as well as the third-party asset owners are resolved.

Where bridges form features of modern infrastructure in densely populated cities and urban areas, the designers will be challenged with site logistics, narrow and busy streets, and third party assets such as railways, underground tunnels, and buried services and it is extremely important in these cases to be more diligent about designing for safety of the temporary works. While the principles can be applied to all bridges, this paper does not cover large suspension bridges on water.

2. Background and UK scenario

Although there were bridge failures in 1950s in USA, Canada and other parts of the world, much of the development in temporary works design in UK was improved when Barton Bridge, Lodden Bridge [2–4] in 1960s and 1970s collapsed during construction and the government was moved to determine whether the industry was in a fit state to manage falsework. Following these major failures in the UK, Professor Stephen Bragg report [4], helped to set the standard for temporary works design and management in the UK. Major failures of UK temporary works have almost disappeared since BS 5975 [5] was published in 1982, following the Bragg report in 1975 into recent falsework disasters in which many lost their lives. In the UK, the design of temporary support trestles would normally comply with the requirements of BS 5975 [6] unless the use of Eurocode 12811 [6] and EN 12812 [7], is stipulated as a contractual requirement. More recently, a Temporary Works forum (TWf) was formed in 2009 [8] in the UK as an independent, non-profit company that operates on a limited cost base. Useful guidance and toolkit and guidance documents are produced by TWf that address the issues of temporary works and applicability of Eurocodes for safer construction in the UK [9]. The general principles of these tool kits generated by the TWf are applicable worldwide.

More recently, the Institutions of Civil Engineers and Institution of Structural Engineers (UK) formed an independent body called SCOSS/CROSS [10] Structural Safety Body in 2005. This Structural-Safety Body is a body devoted to evaluating, and anticipating where possible, trends in the construction industry and issuing warnings where necessary. A confidential (anonymous) reporting system called 'CROSS [10]' broadcasts events which it believes should be more widely known.

While UK standards are developed to a fuller extent to eliminate the catastrophic events in temporary work through the introduction codes, guidance documents [11, 12] we still see that failures occur for various reasons. A publication in ICE Forensic Engineering summarised failures in bridges until 2012 [13] and more than 60 major catastrophic failures across the world have been reported since 2012 to date [14] including the most recent failure in Florida. A few more recent failures are reviewed from the literature survey in this paper to highlight the importance of temporary works.

This paper describes briefly the existing construction techniques of bridges, and reviews more recent failures. There are many types of bridges and but this paper summarises a brief literature survey of current construction methods for precast, prefabricated, and cast-in-situ bridges.

This paper summarises typical scenarios in urban and rural environments and considers a few case histories with a view on how to safely manage the risks associated with the construction of bridges in an urban environment. Most forms of bridge construction, whether preassembled or cast-in-situ will require temporary works on site. These may include, depending on the type of bridge, temporary supports for precast girders or beams, box structures, and temporary staging for cast-in-situ construction of the deck, and may also involve specialist operations for complex forms of bridge, e.g. post-tensioning. Many projects focus on optimising the concrete volumes and steel tonnage but a very few focus on how best to integrate temporary works and buildability concepts into the permanent design at the early stages.

3. Bridge construction techniques: a brief survey

3.1 Precast bridges

Precast concrete members in bridge systems are appealing because they lend themselves well to incorporating Accelerated Bridge Construction (ABC) methods. In some cases, ABC also includes integral column and cap beam systems for bridges utilising precast concrete girders which have several advantages over structures consisting of steel girders or cast-in-place concrete alternatives as shown in **Figure 1** [15, 16].

For single-span bridge decks, temporary support is normally undertaken directly from the abutment bearing shelf or equivalent and this is relatively straightforward. Beams to multi-span bridge decks may require a more complex temporary support system. If the beams are designed in short lengths until they stitched, they will require a temporary support. A common solution is to provide a temporary trestle support either on a temporary foundation or on the permanent foundation of the piers. Beams are landed on the trestle that supports the bridge deck structure until such time that it becomes self-supporting. These trestles can cost significant sums and involve extensive working at height.

Given the consequences of failure and the difficulty of correcting issues that manifest themselves following landing of bridge beams, a robust temporary support structure is a necessity. Factors of safety (as per permissible stress design, e.g. BS 5975) and design factors (as per limit state design, e.g. Eurocodes [17, 18]) can be used amended by the designer to reflect the above. Safety can also be enhanced by building redundancy into the structure. Unless reasonable calculations are made of the specifics, there should be reasonable margins in the design for uneven distribution of load, accidental load and out-of-tolerance assembly.

Figure 1.
Precast girders with GRP deck panels for in situ deck.

In the case of steel structures, and launching steel bridge girders, we may need temporary trestles on temporary foundations. These trestles will require careful consideration due to the launching operations as their stability under dynamic conditions is of utmost importance.

The trestles are generally designed for dead loads, live loads, moving dynamic loads, notional horizontal loads, secondary forces from mis-alignment and will in some cases have impact loads if they are near a live highway.

3.2 Segmental

Segmental construction is one of the most important developments in construction in the last century and is a very well proven method for delivering durable, long span, and repetitive structures that are both cost-effective and visually appealing.

Segmental concrete construction can be executed in two ways: using precast elements or through cast-in-place construction.

The advantages offered by precast elements are mainly related to fabrication, conducted in a plant that produces more consistency in quality products and where segments can be fabricated in parallel with early field construction activities, thus improving scheduling. The main challenges involved with precast segmental construction lie in the logistics and the setup process between the casting yard and the construction site. This includes a large temporary work system involving specialist materials and jacks, and large movable gantry parts.

Alternatively, cast-in-place construction requires that a substructure be completed prior to fabrication of the superstructure. Cast-in-place segmental construction is used when precast segments are too heavy to be shipped or access to the site is too restrictive, which can occur as spans get longer or bridges get wider.

Construction time is a key factor for projects in urban areas which require lane closures, detours, and traffic interruptions to be minimised. Precast concrete segments are often optimal as they can be built and stored until needed for erection, thus reducing the on-site time of large equipment and construction activities, thus increasing the pace of construction. **Figure 2** shows a multi span bridge over a busy highway built using precast and GRP panels for deck.

The choice between precast or cast-in-place primarily depends upon project size, construction schedule, weight of segments, and site access.

Figure 2.
Precast beams and cats in situ deck with GRP panels for formwork over existing highway–4 span bridge.

3.3 Balanced cantilever construction

The balanced cantilever construction method is used when several spans ranging from 50 to 250 m exist. Bridges using this method can be either precast or cast-in-place. Once the piers are built, they are used as an erection platform for precast segments, or to support a form traveller for cast-in-place segments.

This method can also be easily adapted to irregular and long span lengths, congested project sites, rough and water terrain, rail crossings, and environmentally sensitive areas.

The cantilever method is the preferred method for building cable-stayed bridges. Once segments are installed, they are supported by new cable-stays in each erection stage. Since no auxiliary supports are required, it is both an economical and practical solution for long cable-stayed bridges.

As can be seen from the above, all of the techniques will require sophisticated forms of temporary works to enable safe construction. Despite much regulation and improvement in methods/processes in temporary works design, we are still, unfortunately, experiencing catastrophic accidents in bridges (examples include Barton bridge in Manchester in 2016 failure of lifting systems revealed and a 2019 failure in Norway failure of bolts). The failures of temporary works can be grouped into the following main areas;

- Scaffold collapse
- Failure of the lifting equipment, or temporary bearings
- Insufficient design capacity of a cantilevered arm of the cantilevered construction
- Girder and connection failures
- Design and detailing errors
- Incorrect construction sequence
- Negligence and construction errors

If the temporary structures are located close to third party assets, there are further risks that are very expensive to correct there are explained in the following section. The paper covers bridges on land and in populated areas and will not cover large suspension bridges on water.

4. Hazards from and to third party assets

At bridge construction sites in the urban environment, it is not uncommon to come across: party walls, railways, existing grade listed buildings, existing underground utilities and live highways and riverways. The bridge design and construction will have to address how these asset owners can be protected including the safety of public road and railway users where applicable. The requirements of the asset owners may vary but the fundamental aspect of the construction logistics is to ensure that all assets and owners are protected and public safety is ensured. As per the CDM regulations in UK [18], the design and construction should also address life cycle management and also demolition at the end of design life [19].

4.1 Electricity cables and overhead lines

Injuries are usually caused by the explosive effects of arcing current, and by any associated fire or flames that may result, when a live cable is penetrated by a sharp object such as the point of a tool. Such effects can also occur when a cable is crushed severely enough to cause internal contact between the conductors. Injuries are typically severe, potentially fatal, burns to the hands, face and body. There is also a risk of electric shock.

Inadvertent contact or being in close proximity to overhead electricity lines with equipment such as scaffold tubes, irrigation pipes, metal ladders or vehicles, such as cranes, poses a risk of electric shock. Direct contact with overhead lines is not necessary as electrical current can arc or flashover any gap between the overhead lines and the object. It is therefore important to ensure sufficient clearances for overhead lines are maintained and plant/surcharge loading is restricted from trestle foundations.

4.2 Gas pipes

Damage to underground gas pipes can cause leaks that immediate or time related that may lead to fire or explosion. The ground pressure from trestles or crane mats or outrigger could pose a risk to the stability of the gas pipes and therefore a risk assessment is required from the surcharge loading.

4.3 Water pipes, sewers and drains

Although damage to water pipes is less likely to result in injury, the following may occur;

- A jet of water from a main can injure a person.
- Leaks of water from underground pipes can affect adjacent services and reduce support for other structures.
- Damage to, or removal of thrust blocks can result in sudden loss of containment and the movement of pipe fittings that may travel some distance or cause impact damage. While some sewage is pumped at pressure, sewers are generally gravity-fed, and the main hazards from damage to a sewer are the possibilities of contamination and subsidence.

The type of materials for these water/sewage pipes, will include brick, cast iron, ductile iron, clay or concrete. Some asset owners, will give limitations on the surcharge loading and acceptance criterion for ground settlements or strain, an example of which is shown below from Thames water UK [20]. Current regulations in UK will require a risk assessment before placing any temporary foundations or surcharge on these assets (**Tables 1** and **2**).

4.4 Telecommunications cables, broadband and fibre optics

Damage to telecommunication and TV cables may require expensive repairs and can cause considerable disruption to those relying on the system, especially emergency or essential services. The risks of direct personal injury are normally low, but claims for consequential losses may be substantial.

PIPE TYPE	Diameter (mm)	Allowable Increase in Strain (□□)		Rotation (deg.)
		Tension	Compression	
Brick Sewer (red / yellow / blue brick)	N/A	500	25% of the allowable stress	N/A
Cast Iron Lead-yarn joints	N/A	100	1200	0.1
Ductile Iron (Lead-yarn gasket joints)	N/A	500	700	0.5
Ductile Iron (Rubber gasket joints)	N/A	500	700	2.0
Steel	N/A	450	450	1.5
Vitrified Clay	<125	80	400	0.5
	>125	80	400	See Table 2
Concrete (unreinforced)	<225	20	400	0.5
	225 – 750	40	400	See Table 2
	>750	60	400	

Table 1.
Assessment criteria for existing Thames Water pipeline and sewer assets (reproduced from Ref. [20]).

Diameter (mm)	Rotation (deg.)
< 375	2.0
375 – 750	1.0
750 – 1400	0.5
> 1400	0.3

Table 2.
Maximum rotation for vitrified clay and concrete pipes.

It is important to ensure that surcharge loading from temporary foundation on buried cables is limited.

In addition to the above utility services, the presence of other pipes and cables should be anticipated. These include fuel oil pipes at States housing developments and private/security electricity and telecommunications cables. Risk assessment will be required on a case by case basis to suit to owners' requirement.

4.5 Underground or above ground railway or cable tunnels

Any movement of tunnels if they are underground or above ground (cut and cover) is very critical for the performance and therefore the asset owners will have very stringent requirements for the movement temporary works plant around these structures. Generally clearances need to be maintained between the plant and the assets and this will vary depending on the depth of the asset below the ground. No plant can operate within these zones while constructing the bridges. An example of the Cross Rail exclusion zone in London [20] is shown in **Figure 3**.

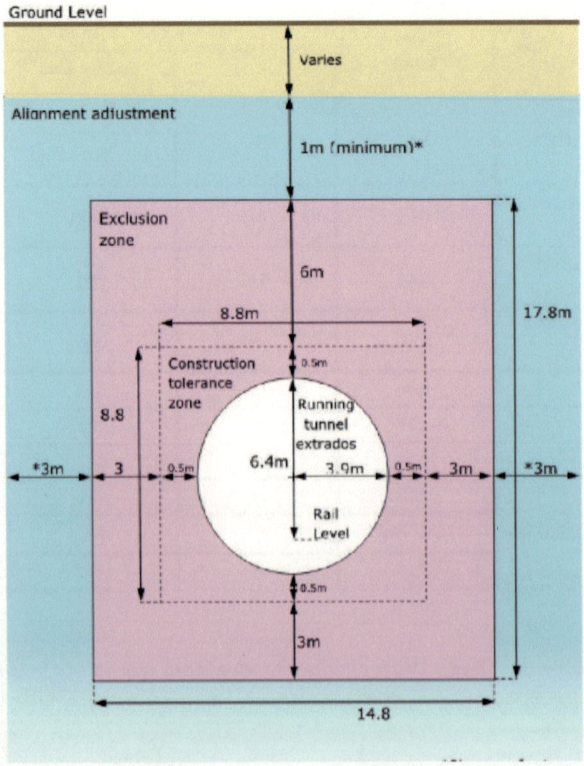

Figure 3.
Exclusion zone limits for crossrail underground lines (figure reproduced from Ref. [21]).

4.6 Party walls or other grade listed buildings

These include third party buildings, existing walls of private owners, and grade listed buildings. Generally, if the temporary works for bridge construction works are carried out by the contractor near these assets, he will have to ensure that the damage to either Partywalls or nearby listed grade buildings is minimal. In UK, there are party wall agreements on movements and crackwidths as per Boscardin guidelines [22].

In order to manage the risks from bridge construction operations, it is necessary to prepare a carefully coordinated construction method statement with all the supply chain involved and which includes the sequence, hazard identification, and a risk assessment. A management procedure is required to manage the residual risks on site and contingency measures to mitigate or control the risks.

5. Dealing with hazards: construction method statements, and risk assessments

Catastrophic events in construction are real issues which require proper consideration by all stakeholders, led by directors and senior staff.

These potentially catastrophic events are sometimes referred to as 'Top Events'. It is appreciated that they can have a disastrous impact on a company's reputation and well-being and upon society. The process of examining the risk of a catastrophic event requires that a 'safety case' is prepared, based upon a safety risk assessment.

CIRIA and HSE UK reports [23, 24] have looked at the risks of 'Catastrophic Events' in the UK construction industry and summarised its findings in the report. This report identified the types of events, reason for occurrence, and control measures and included:

- The types of catastrophic event which have occurred or which might occur during construction
- The reasons for occurrence when there have been (or could have been) catastrophic events during construction, including an examination of the underlying factors
- The controls which would contribute to an avoidance of a catastrophic event
- Where the UK construction industry could improve

It was clear that there have been Catastrophic Events with major consequences. Their importance was recognised by the industry, although it is considered that in their day-to-day work few people realised the severity of what might happen if things went seriously wrong.

The key issues proposed in this reports are as follows:

- Issue 1: The industry should recognise that catastrophic events need further attention
- Issue 2: Corporate risk management systems should be improved
- Issue 3: Knowledge, skills and experience of safety risk management should be raised
- Issue 4: Communication and interface management should be improved
- Issue 5: Competence is key
- Issue 6: Effective management of temporary works is crucial to success
- Issue 7: Independent reviews should be employed
- Issue 8: The industry should learn from experience.

Method statements and drawings need to fully detail all aspects of the works. All safety critical items and hold points/permits to load should be identified.

- Low likelihood/high severity items are to be given careful consideration with appropriate monitoring at all levels.
- More invasive questioning and understanding of sub-contractor's Method Statements and Risk Assessments

5.1 Risk assessments

The terms used in risk assessment vary considerably in the literature. In this report, a harm is defined as an adverse effect on a person. It might be, for example, a serious illness or injury, but effects on well-being are also taken into account. A hazard is a potential cause of harm to a person, for example a faulty staircase. A hazardous situation exists when a person is exposed to a hazard, for example by using a faulty staircase. The risk associated with a hazard is a function of (a) the likelihood of the hazard causing harm and (b) the severity of the harms or their consequences.

Safe systems of work shall comprise:

- Assess the risks.
- Plan the work - obtain all information relating to the work carried out.
- Check that the plans are accurate.
- Carry out the work in a safe manner.
- At all stages THINK and REVIEW.

The assessment of risk should be considered at all stages of the work, from planning through to final reinstatement. This may be accomplished by the use of formal risk assessments coupled, where necessary, with a work permit system. Risk assessment should include all related work activities and identify training and competency needs as well as the level of supervision required for the risks involved [10].

6. Need for construction: pre trials, tests and other monitoring

Sometimes, it is necessary to carry out trials on site before the bridge structure construction begins. This may take include some mock structures to be built on site to validate all the assumptions in the design. The purpose is to ensure that the sequence of construction can be carried out safely. Examples include test loading on the trestles or surcharge loading, and monitoring of displacements etc. and mock assembly in case of structural steel joints. Eurocodes can be implemented to optimise the temporary works design if suitable testing is carried out.

In the following section, we give more recent literature survey of failures in the temporary works that involved public and third party asset owners and give additional examples of how we have managed the risks (as explained in Sections 3 and 4) during the construction for the projects in which Sir Robert McAlpine and their joint partners were involved. Our recent project also demonstrate how we have used trials on site for back analysis and how monitoring was carried out to improve the confidence in predictions and stakeholder assurances.

7. Survey of lessons learnt from recent failures

7.1 Barton Bridge, Eccles, Feb 1959 & 2016

In the UK the First Barton bridge collapsed in 1959 whilst erecting 4 No, 200 ton steel girders, 80 ft. above the ground [3, 14]. The supporting scaffolding collapsed bringing down the girders and killing 4 men. Sixty men, that would normally have been on the girders, were lining up for their pay at the time. Ironically after 57 years, at the same location in 2016, temporary lifting system failed collapsing a major chunk of the new bridge span across the river as shown in **Figure 4a** and **b**.

7.2 GE 19 bridge on East London Line

GE19 is an 84 m long, single span Warren truss girder bridge on a 3.3% gradient from East to West with a bridge weighed 1300 tonnes post launch and had an In-situ deck on 'Omnia' permanent formwork [25]. Minimum clearance of 650 mm to overhead lines. It was found that the bridge had moved longitudinally 38 mm out of position post launch. This necessitated corrective plan jacking that had not been envisaged during pre-planning. At approximately 19:15 hours on 28th May 2008, an hour after work had stopped, the site security guard heard three loud bangs. The Bridge deck had dropped by approx. 200 mm resulting in damage to scaffolding and bearings. Five planks fell onto the live rail below the bridge, ponded water onto the overhead lines and track and the track had to be closed. There were no injuries to any members of the public or employees. PTFE material had been placed on a slope,

(a) (b)

Figure 4.
(a) Barton bridge scaffold collapse (1959) and (b) lifting failure of a modern bridge (2016) (figures reproduced from Refs. [3, 14]).

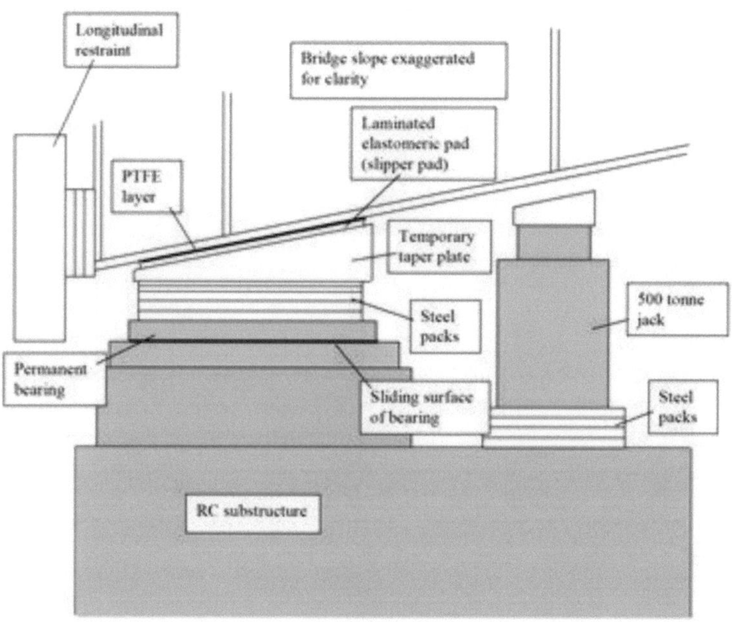

Figure 5.
Temporary bearing.

the vertical load of the bridge generated a horizontal force component. The presence of a second slip plane allowed the wedge of packing between the PTFE surface and bearing surface to be ejected as in **Figure 5**. A similar incident did not occur at the West abutment because the permanent bearings were of a fixed type and hence did not fail (see **Figure 6**).

7.3 Motorway bridge temporary works collapse Colombian Bridge

In March 2019 the temporary works of a motorway bridge near Ancona, Italy, shown in **Figure 7** [26] failed with the immediate collapse of the bridge deck onto cars passing below sadly killing two members of the public [14, 26]. The motorway was subsequently closed. An example of both the human and financial consequences of errors in construction. A design error was blamed for the collapse of the bridge.

7.4 Grayston Drive collapse (Johannesburg)

Two people were killed and 19 injured when the formwork supporting the under-construction bridge collapsed unexpectedly in 2016 shown in **Figure 8** [27]. The initial inquiry showed that some of the site inspection registers for the period just before the collapse were not available [27].

Figure 6.
Bridge launch over railway (reproduced from Ref. [25]).

Figure 7.
Colombian bridge failure (reproduced from Ref. [26]).

7.5 Norway Bridge Collapse 2018

A Hydrogen embrittlement crack has been identified as causing the failure of one of the cable anchor bolts of the Norway Halogaland Bridge shortly after installation shown in **Figure 9** [28]. The initial enquiry showed that that the root cause of the cracking is the hydrogen exposure of the bolts. It is not known if the bolts were exposed to Hydrogen during manufacturing, transportation or at the site.

7.6 Injaka Bridge, South Africa 1998

Incorrect positioning of temporary bearings during incremental launching was identified as the primary cause of the fatal 1998 Injaka Bridge collapse in South Africa [29].

Inexperienced design and construction staff, poor construction quality control and a failure to react to a 'clear warning that all was not well' with the structure, led to the disaster. At 300 m long, 14 m wide and up to 37 m above the river bed, Injaka Bridge was a major structure and the consultant and contractor had extensive experience with such incrementally launched post-tensioned structures. The collapse occurred after the contractor had slid out five of the 20, 15 m long sections of the 3 m deep box section deck. The sixth segment was being jacked as the structure collapsed. At that point the concrete deck extended 24.4 m beyond pier 2 with the leading edge of the 27 m long launching nose projecting 7.1 m beyond pier 3.

Figure 8.
Collapse of temporary supports 2018 (figure reproduced from Ref. [27]).

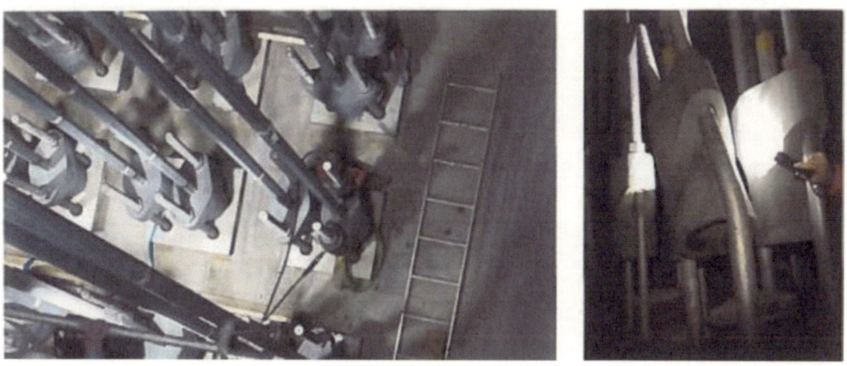

Figure 9.
Hydrogen cracking in bolts (figure reproduced from Ref. [28]).

The primary cause of the collapse was found to have been the positioning of temporary bearings on which the deck structure slid out during construction as shown in **Figure 10**.

Figure 10.
Failure of a multi span bridge (figure reproduced from Ref. [29]).

Figure 11.
Florida Bridge collapse, 2019 (reproduced from Ref. [25]).

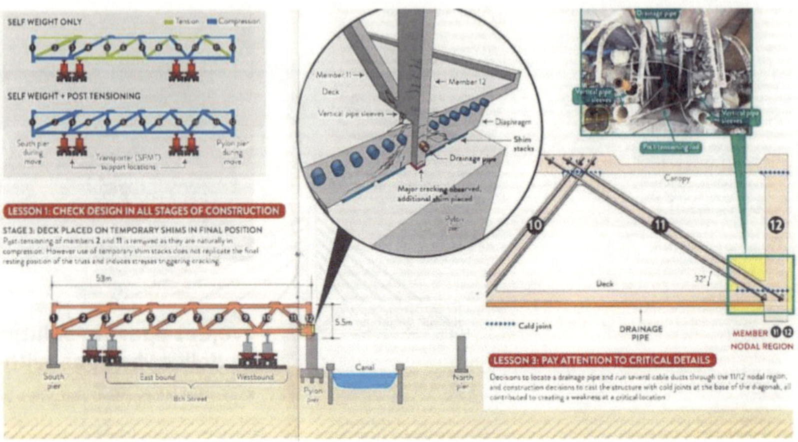

Figure 12.
Node 11 failure (reproduced from Ref. [30]).

7.7 Florida Bridge collapse 2019

The initial investigation showed that where the tow truss members meet, at joint 11, there was overestimation of the capacity and under estimation of the loads as shown in **Figures 11** and **12** [25, 30, 31].

8. Case studies of mitigation

Examples of case studies from our recent project experience published by the author in various reports (internal and external) are outlined below. These case studies show how the temporary works design should address the stake holder's expectations in managing risks. Five case histories discussed here show the importance of design, checking, and monitoring and back analysis.

8.1 Case Study 1: Arsenal bridges, London, UK

An example of assets affected by the bridge launch—temporary trestle foundations next to network rail assets, and piling plant and movement of construction plant on the sensitive underground railway tunnels (**Figure 13**).

Assessments were carried out to ensure that the settlements, and movements and surcharge loadings were managed. Monitoring was carried out to gain confidence in the predictions and to ensure that mitigation measures can be placed (**Figure 14**).

8.2 Case Study 2: Gogarburn Bridge, Scotland

The road bridge is on the new spur off the A8 into the new headquarters for the Royal Bank of Scotland. The bridge deck was being erected on temporary trestles, located on either side of the road and in the central reservation. The box girders will be lifted into place in half span sections, each weighing approximately 73 T. Secondary beams span between the box girders to support the deck. The deck was cast on 'omnia plank' permanent formwork. The sequence is depicted in **Figure 15** and the temporary trestles that required design for impact loads are shown in **Figure 16**.

Figure 13.
Bridge launch.

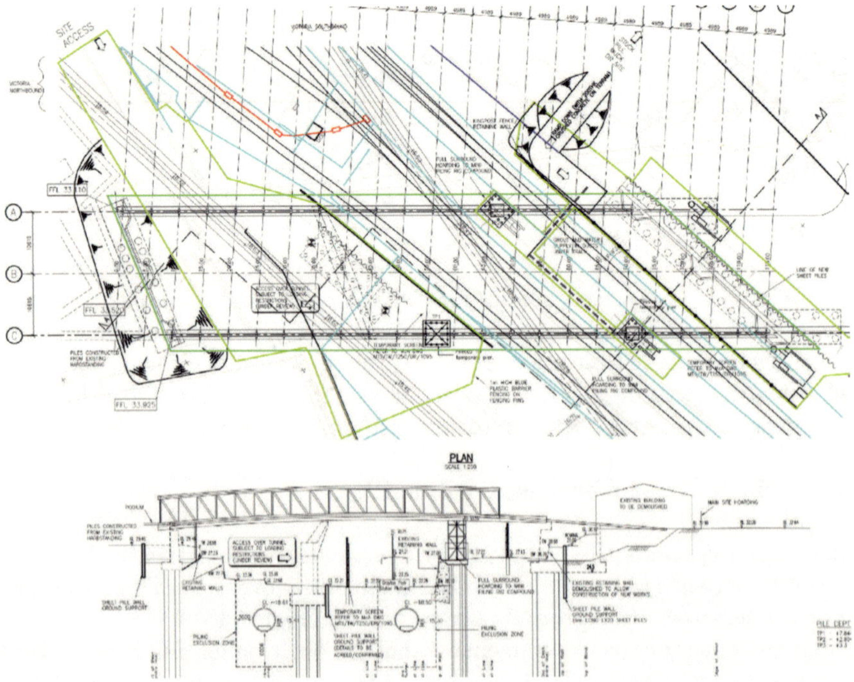

Figure 14.
General arrangement plan and section showing trestles, UG services, tunnels, and network rail.

Figure 15.
Erection sequence of the bridge, trestle.

The arch was being erected in three sections. Each of the two outer pieces was supported by the bearing plate at one end and a trestle on top of the deck at the other.

Finally the centre section was dropped in to complete the arch. The arch was welded together, 25 m above the road. Ten tension rods support the bridge in the centre.

Figure 16.
Temporary trestles near live highway.

8.3 Case Study 3: Motorway M74, UK, Bridge Launch from the temporary trestles

As a part of the M74 completion project in Glasgow, Sir Robert McAlpine Design group, working closely with the site based Joint Venture, has been responsible for many elements of design and checking for the West section of the work [31]. Included in the West section is a 1350 m length of an elevated structure formed of trapezoidal steel girders with in-situ concrete deck and parapets. This section passes over three sections of the overground line which are a part of Network Rails assets; the Paisley Line, the Cook Street Link, and the West Coast Mainline. Parts are also close to a section of the Strathclyde Passenger Transport (SPT) underground line.

With the exception of the West Coast Mainline section, the steel sections were lifted in place using mobile cranes. These sections were supported on temporary trestles whilst they were welded. The section of superstructure spanning over a 65 m wide cutting containing the West Coast mainline was installed by launching the bridge as shown in **Figure 17a**. A 235 m long, 4200 tonne section was formed to the West of the cutting, complete with a section of in situ deck and parapet, and then launched a total of 166 m during a series of short night-time possessions. The

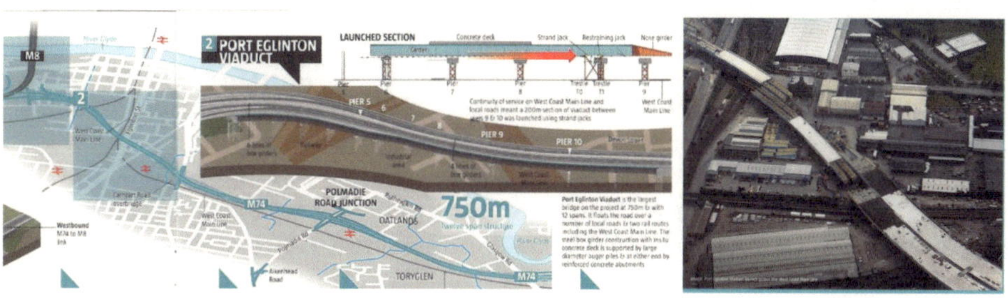

(a) (b)

Figure 17.
(a) Launch of the bridge over the live highway and railway, (b) aerial photo of the launch (extract from Ref. [31]).

cantilever distance to the first temporary trestle support beyond the cutting is 89 m. In order to ensure safety to both Public and the existing network rail assets, the temporary works design included the following and was developed collaboratively with the supply chain partners in the joint venture project.

i. Assessment of the ground was necessary for designing working platforms for all the plant and craneage required for the steel lifts. Outrigger loads from the cranes were up to 4200 kN, with 6000 kN of superlift loadings.

ii. Assessing the effect of the applied loads on the Scotland Public Transport (SPT) infrastructure.

iii. Design of frames and anchored rock netting to protect Network Rail assets during piling and pier construction operations adjacent to the railway cutting.

iv. Design of temporary support trestles and their support piles to take the vertical and horizontal loads applied during the bridge launch operations. The calculated vertical loads were up to 18,000 kN on a single temporary leg under the worst case loading. The design of the trestles evolved as work proceeded as exploratory works were carried out in advance of the piling and services or obstructions were encountered:
One pair of temporary trestle bases was cast adjacent to a Network Rail Retaining and within a main road. The design had to take account of the need to avoid unacceptable movements to the nearby masonry retaining wall.

v. The six 1200 mm piles to each base had to be installed in locations which avoided impacting on three gas mains, electric cables and a brick sewer.

vi. Design of additional support walls to take the bridge launch loads on the pile caps under the temporary loading conditions.

Working collaboratively with our supply chain partners, loads on each of the trestles for the launching operations were derived and were used in the analysis of temporary trestles. Trestles foundations included piles where space was restricted and pad foundation where space was not restricted. The analysis for various combinations of normal loads and accidental loads were carried out in ANSYS and typical results from the analysis were summarised in **Table 3**. These were used for monitoring the loads and the movements of the foundations using targets on trestles and foundations.

8.4 Case Study 4: construction of a segmental arch bridge over a railway, Dobwalls Bypass, UK

This case study, published in detail in Ref. 32 by the author, shows how the temporary works were managed by the project team in order to ensure that there was no repeat of 'Gerrards cross failure'. The failure at Gerrards Cross, reported in 2005, demonstrated the importance of controlling the backfill and carefully controlling and monitoring the deflections of the arch bridges during construction. The 87 m long arch unit spanning 15.5 m with a rise of 5.6 m was built on the monolithic principle, which means it acts as one single structure. The radius of the curved track underneath is 500 m, which proved a challenging aspect of the project.

Item	Trestle 1	Trestle 2	Trestle 3
Vertical settlement	8 mm	8 mm	0 mm (under lifting strut). 6 mm elsewhere.
Anticipated lateral movement (at pile cap). Parallel to launch	10 mm	10 mm	9 mm
Maximum lateral movement (worst case soils) at pile cap. Parallel to launch	—	—	21 mm
Maximum lateral movement (worst case) at pile cap. Perpendicular to NR retaining wall	—	—	14 mm
Worst case lateral movement at network rail wall for worst case soils and bridge being pulled back (westwards)	—	—	3 to 5 mm
Total maximum vertical load on support	+5000 kN	34,500 kN	20,000 kN
Total minimum vertical load on support	−4350 kN	—	—

Table 3.
Amber trigger levels for movements and loads on trestles.

The structure was in the form of a short tunnel, a proprietary system by Asset International, comprising pre-cast concrete arch segments springing from an RC slab and upstand, supported on piled foundations. The arch was formed from pre-cast concrete elements using a proprietary arch system with an elliptical cross-section. The two ends of the tunnel consisted of portal sections which are bevelled to follow the slope of the new embankment for the realigned section of road. Since the bridge structure is made up of arch-shaped pre-cast elements, the elements at the ends of the tunnel are cut-off, and no longer form a full arch. These truncated elements, therefore, are connected to each other by means of cast-in-situ reinforced concrete collars to form a monolithic reinforced concrete shell. This monolithic shell includes the two outermost full arch rings which allow all loads acting on one side of the bevel to be transferred to the other side and to the foundations. Working collaboratively with the project team, we have developed a safe system of construction as described below.

In order to predict the behaviour of the tunnels during construction and to advise the construction teams on the methodology, it was necessary to model each stage both in 2D and 3D models. The construction sequence was represented in the analysis by a total of 15 stages as shown in **Figure 18**. Full details of the FE model, assumptions, approach, and the sequence including sensitivity analyses were published in a separate paper Ref. [32] by the author(s). Only extracts from Ref. [32] are presented in this paper.

The construction phases are summarised below.

Phase 1: Establish initial conditions for existing ground and railway embankment;

Phase 2: Place new fill up to the level of the existing railway embankment, and install piles and pile caps;

Phase 3: Construct arch;

Phase 4: Place fill for the new road embankment away from the arch structure up to 75% of the overall height of the arch segments;

Phases 5–13: Place backfill against the arch with a maximum differential between sides of 600 mm, with compaction load of 11.5 kN/m^2 on the surface of the fill;

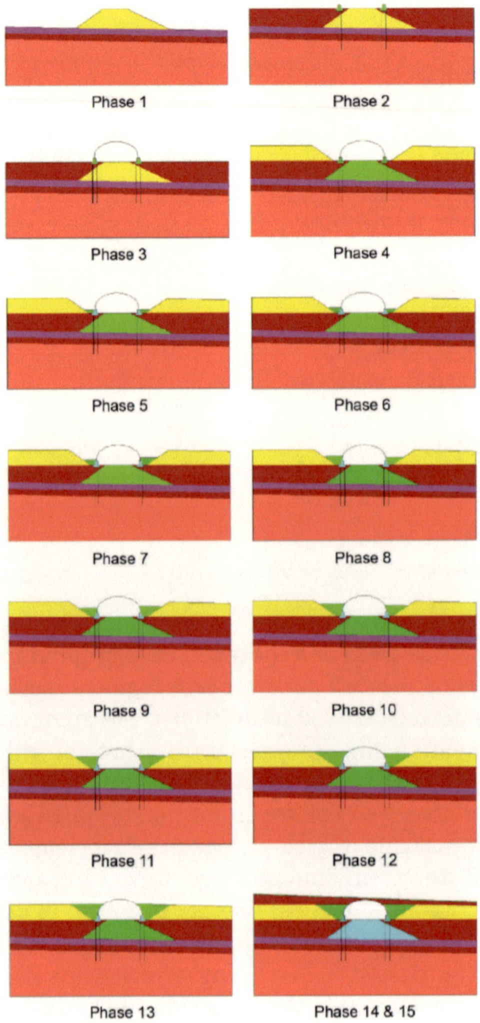

Figure 18.
Construction sequence model analysed in the finite difference software (FLAC).

Phase 14: Place final layer of fill over the arch, with compaction load of 11.5 kN/m^2 over the full width of the model on the top surface;

Phase 15: Remove compaction load of 11.5 kN/m^2 on top surface of the model.

It was important to carry out sensitivity studies with respect to soil stiffness, backfill characteristics, interface stiffness, and initial conditions of the arch to establish lower and upper bounds of movements. These sensitivity studies helped us to develop a safe and robust scheme of backfilling sequence and helped us to set new trigger limits for safe construction. Closed-form solutions predict the movements for a fully backfilled scenario however they will not predict the movements for unsymmetrical backfilling on either side of the arches and therefore numerical models in 2D and 3D will give insight to real behaviour. The models developed here in 2D and 3D, therefore, gave insight into the development of the movement throughout the construction process. The power of the modelling is demonstrated by comparing the analytical results with observations on site. The site observations matched well with the numerical predictions from 2D and 3D models as shown in **Figure 19**.

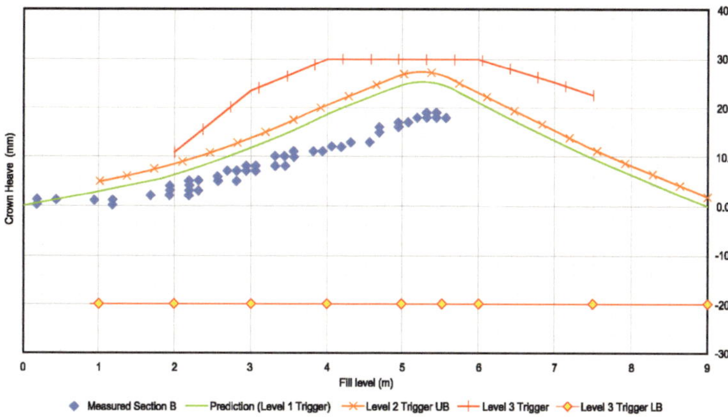

Figure 19.
Movement of the crown of the arch while backfilling sequence is progressed to build the highway bridge over the railway.

8.5 Case Study 5: construction of a hanging building from the truss over railway, Bull Ring, Birmingham, UK

The Bull Ring Redevelopment in Birmingham consisted of demolishing the existing 1960s concrete shopping centre and replacing it with a new one. The northern part of this new complex lies directly above the New Street South railway tunnels, which carry the main lines to London and the West Country through them. During the redevelopment work there was the potential to affect the railway tunnels at various stages of construction.

To maximise the available retail space at the northern end of the development, 2 No. hanging structures were to be constructed to extend the development over the Northern Arm road, with the pedestrian footbridge described above providing the 'sandwich' between the hanging structures. To support these hanging structures a structural steelwork bowstring truss is positioned at either end of each structure, spanning across the Northern Arm to carry all other intermediate steelwork, reinforced concrete floor slabs, roof, cladding, etc.

The western hanging structure is supported by trusses T1 and T2, which are supported at their northern end by double columns supported off a reinforced concrete pile cap founded on a cluster of mini piles constructed within the basement of the Rotunda. At its southern end these trusses are again supported by a twin steel column section founded behind the contiguous piled wall and thus forming part of the main development structural frame. The brick arch railway tunnels are not continuous for the full length of the Northern Arm. At its western end the brick arch tunnels give way to a reinforced concrete road bridge, built 1961–1962. This road bridge continues in a westerly direction towards the New Street station junction, noting that immediately northwest of truss T1 the road bridge deck slab is discontinuous with an open section of railway exposed and only protected by a 1.8 m high concrete parapet wall constructed around the opening. Working collaboratively with our supply chain partners, we have developed a safe erection methodology as described below.

Once the fabrication location was established, crane sizes and locations were firmed up, thus allowing a detailed crane analysis to be carried out to produce theoretical outrigger loads, including any redistribution of loads resulting from any crane slewing, jibbing in/out, etc. during the lifts. The Railtrack structures were then assessed under these loads, with feedback to the site team accordingly if the

Figure 20.
Erection of 140 T truss over the railway with a soil cover of 3 m.

structures were likely to be overstressed, with recommendations to relocate cranes, increase outrigger mats, distribute outrigger loads onto twin mats, etc.

Once craneage locations were finalised, craneage layouts were firmed up, structural checks were completed, scaffold layouts finalised and method statements produced. Craneage used for the fabrication of the trusses and associated erection of the temporary works support scaffold varied from truss to truss, but for each truss one of the cranes used for the tandem lift of that truss was utilised as the service crane for the above works. Therefore, for trusses T1 and T2 a 400 T crane was used with an 800 T crane used for trusses T3 and T4. As part of the checks undertaken during the initial piling works to establish constraints for the piling plant, its location, and associated excavations, a considerable amount of design checks using finite element analysis had been undertaken on the brick arch tunnels. To satisfy railtrack/network rail requirements the following monitoring equipment had been installed into the tunnels before works started and included, electro level beam surveys, vibration sensors, tilt meters, and tape extensometers.

Temporary works design checks undertaken were based on theoretical outrigger loads prepared by the crane manufacturers following assessment of the different lifts by the different cranes. On the basis that these outrigger loads did not cause distress to the structures below, it was essential that outrigger loads were checked to ensure that the theoretical maximum loads were not exceeded. Hence use was made of the crane digital outrigger load readout indicator to monitor these loads. The details of the finite element analysis and the assumptions are presented in a full paper in Ref. [33] and only extracts from Ref. 33 are presented in this paper.

During each lift and at various times during the fabrication of the trusses, the tunnel monitoring system PC was attended full time by one of our site engineers to observe any changes, notably deformation movements. A site engineer would also monitor the full time the crane outrigger load indicator during major lifts.

During the crane lifts and during the fabrication of the trusses no discernible deformation to the tunnel was recorded—noting that, when the tunnel monitoring system PC was not being observed full time, the system did activate a telephone alarm once a deformation of 7 mm occurred.

The crane outrigger loads, as observed on the outrigger load indicator, were generally well inside the theoretical figures. On truss T1 lift the 800 T crane on the bridge beams had a maximum outrigger load of 74 T against a predicted value of 106 T. On truss T2 lift the 800 T crane on the tunnel central wall had a maximum outrigger load of 92 T against a predicted value of 98 T.

One the same lift the 400 T crane on the transfer beams spanning across the beams had a maximum outrigger load of 99 T against a predicted value of 120 T. **Figure 20** shows the truss T1 over the tunnels.

Immediately following each lift on the first available track possession/isolation a visual inspection of the tunnel was undertaken and no visible signs of distress and

Figure 21.
Non-linear finite element analysis of soil structure interaction for the loads from erection of the truss over the live railway.

	North Tunnel				
	Non-linear analysis		Linear analysis		Observed
	Lateral	Vertical	Lateral	Vertical	Vertical
A	-0.35	-0.68	-0.24	-0.65	-0.7
D	0.05	-1.25	0.29	-1.20	-1.01
G	0.46	-0.64	0.52	-0.66	-0.7
	South Tunnel				
I	-0.52	-0.35	-0.39	-0.47	-0.9
K	-0.18	-0.98	-0.04	-1.13	-1.3
M	-0.24	-0.37	0.32	-0.49	-0.4

Table 4.
Measured and observed movements of the arch for 70 T rig on the tunnels.

movement were ever observed. Tape extensometer checks were also undertaken and again no significant movement attributable to the crane lifts was ever recorded.

Finite element models were developed in ANSYS (commercially available software) in view of the above. In particular, two-dimensional linear and non-linear global analysis and a three-dimensional non-linear local analysis were carried out. **Figure 21** shows the finite element model. Planar elements were used for two-dimensional analysis whereas solid elements with no tension were used in the three-dimensional analysis.

ANSYS results were calibrated against site observations by making test trails on-site by surcharging pile plant loading on the tunnels and results are shown in **Table 4**.

Analysis method, assumptions, and further details of the finite elements are presented in the detailed paper by the authors in CIRIA report on tunnelling [33]. Only extracts are presented in this paper.

9. Summary

Temporary works are an integral part of the safe construction of bridges and should be designed by competent bodies. Independent checks, and balances are to be in place to ensure that the public and asset owners are protected by the construction method, construction sequence of bridges. This paper summarises more recent failures and draws the conclusion that some of the best practice guidelines as developed in the UK can be adopted outside the UK. Early engagement by the contractor and communication to all the parties involved play a significant role in the safe delivery of the works at the site. Every aspect of the temporary work should be looked into in detail with reasonable margins of safety. Regular monitoring during the construction to ensure that the early warnings are not exceeded and are vital to verify the performance is as predicted and that no unsafe condition is being approached.

Acknowledgements

The author is thankful to Parsons Brickenhoff for supply of the data and Asset International for sharing site measurements and photos. The author is thankful to our JV team of M74 and Paul Doughty of Sir Robert McAlpine Design Group for sharing data and photographs.

Conflict of interest

The authors declare no conflict of interest.

Author details

Ganga Kasi V. Prakhya
Sir Robert McAlpine Ltd, Eaton Court, United Kingdom

*Address all correspondence to: g.prakhya@srm.com

References

[1] Grant M, Pallet P. Temporary Works: Principles of Design and Construction. London, UK: ICE Publishing; 2012

[2] Bridle R, Sims F. The effect of bridge failures on UK technical policy and practice. Proceedings of the Institution of Civil Engineers: Engineering History and Heritage. 2009;**162**(1):39-49

[3] Smith D. Bridge failures. Proceedings of the Institution of Civil Engineers: Part. 1976;**1**(60):367-382

[4] HSE (Health and Safety Executive). Final Report of the Advisory Committee on Falsework. The Bragg Report. London, UK: HMSO; 1976

[5] BSI. BS 5975: 2008+A1:2011. Code of Practice for Falsework. London, UK: BSI Publication; 1982

[6] BS EN 12811-1. Temporary Works Equipment. Scaffolds. Performance Requirements and General Design; 2003

[7] BSI. BS EN 12812: 2008. Falsework. Performance Requirements and General Design. London, UK: BSI; 2008

[8] Temporary works forum. Toolkits for Design. Available from: https://www.twforum.org.uk/home [Accessed: January 2020]

[9] Hewlett W, Jones A, Marchand S, Bell B. Re-visiting Bragg to keep UK's temporary works safe under EuroNorms. Proceedings of the ICE: Forensic Engineering. 2014;**167**(2):58-68

[10] Structural Safety Reports on Temp Works. Available from: https://www.structural-safety.org/ [Accessed: January 2020]

[11] Soane A. Temporary works toolkit. Part 5: Temporary works failures—What are the common causes? The Structural Engineer. 2017;**95**(1):24-26

[12] Bennion C. Temporary works associated with precast concrete bridge beam construction. In: Proceedings of the Institution of Structural Engineers, London, May 20. 1987. pp. 37-41

[13] André J, Beale R, Baptista AM. A Survey of Failures of Bridge Falsework Systems Since 1970. ICE—Forensic Engineering, Institute of Civil Engineers; 2012. 4/12

[14] Wilkipedia—Bridge Failures. Available from: https://en.wikipedia.org/wiki/List_of_bridge_failures [Accessed: January 2020]

[15] Broughton K, Gill J. Bridge Installation Techniques. London: ICE Publication; 2015

[16] Khan MA. Accelerated Bridge Construction, Best Practices and Techniques. Elsevier Publication; 2015. ISBN: 97820-12-407224-4

[17] BSI. BS EN 1990: 2002 + A1: 2005. Eurocode—Basis of Structural Design (Incorporating Corrigendum December 2008 and April 2010). London: BSI; 2005

[18] BSI. BS EN 1991-1-6: 2005. Eurocode 1. Actions on Structures. General Actions. Actions during Execution. London, UK: BSI; 2005

[19] HSE Publication. Construction (Design and Management) Regulations (CDM). London: HSE Publication; 2015. Available from: https://www.hse.gov.uk/construction/cdm/2015/index.htm [Accessed: January 2020]

[20] Thames Water Publication. Guidance on Piling, Heavy Loads, Excavation, Tunneling and Dewatering. Available from: https://developers.thameswater.co.uk/Developing-a-large-site/Planning-your-development/

Working-near-or-diverting-our-pipes [Accessed: January 2020]

[21] Cross rail Information for developers. March 2019. Available from: http://www.crossrail.co.uk/route/safeguarding/?folder=/10/856/asset/6043 [Accessed: January 2020]

[22] Gaba A, Hardy S, Doughty L, Powrie W, Selemetas D. Guidance on embedded retaining wall design. In: CIRIA Technical Report C 760D. London: CIRIA Publications; 2017

[23] Gilbertson A, Kappia L, Bosher L, Gibb A. Guidance on catastrophic events in construction. In: CIRIA Publication C 699. London, UK; 2011

[24] Preventing catastrophic events in construction. In: Health & Safety Executive Research Report, HSE RR 834. London: HSE Publication; 2011

[25] Bridge Design & Engineering. 2019;**95**(2nd Q):6

[26] Bridge Design & Engineering. 2018;**91**:6

[27] Bridge Design & Engineering. 2018;**92**:6-7

[28] Bridge Design & Engineering. 2019;**97**:7

[29] New Civil Engineer. UK: Publication of Inst of Civil Engineers; August 2002

[30] New Civil Engineer. UK: Publication of Inst of Civil Engineers; January 2020:38-39

[31] New Civil Engineer. Supplement on M74. UK: Publication of Inst of Civil Engineers; June 2011

[32] Prakhya G, Hopkin I, Hansford B. Construction of a concrete segmental arch bridge over a railway. Proceedings of the Institution of Civil Engineers—Bridge Engineering. 2019;**172**(3):226-240. DOI: 10.1680/jbren.18.00027

[33] McKibbins L, Elmer R, Roberts K. Tunnels: Inspection, assessment and maintenance. In: CIRIA Technical Report C671. London: CIRIA Publication; January 2010

Chapter 7

Geometric Accuracy of Digital Twins for Structural Health Monitoring

Ruodan Lu, Chris Rausch, Marzia Bolpagni, Ioannis Brilakis and Carl T. Haas

Abstract

We present an exploratory analysis of the geometric accuracy of digital twins generated for existing infrastructure using point clouds. The Level of Geometric Accuracy is a vital specification to measure the twinning quality of the resulting twins. However, there is a lack of a clear definition of the Level of Geometric Accuracy for twins generated in the operation and maintenance stage, especially for structural health monitoring purposes. We critically review existing industry applications and twinning methods. To highlight the technical challenges with creating high-fidelity digital replicas, we present a case study of twinning a bridge using real-world point clouds. We do not provide conclusive methods or results but envisage potential twinning strategies to achieve the desired geometry accuracy. This chapter aims to inform the future development of a geometric accuracy-based evaluation system for use in twinning and updating processes. Since a major barrier for a fully automated twinning workflow is the lack of rigorous interpretation of 'geometric accuracy' outside design environments, it is imperative to develop comprehensive standards to guide practitioners and researchers in order to achieve model certainty. As such, this chapter also aims to educate all stakeholders in order to minimise risk when drafting contracts and exchanging digital deliverables.

Keywords: digital twin, geometric accuracy, point clouds, bridge, structural health monitoring

1. Introduction

In the wake of the Notre Dame Cathedral fire, digital scans collected by Dr. Andrew Tallon [1] offer the hope for future restoration. One question raised is, what Level of Geometric Accuracy (LOGA) can the reconstructed digital replica achieve with respect to the physical asset? In the Architecture, Engineering and Construction (AEC) sector, operation and maintenance (O&M) costs can range between 60 and 80% of total life cycle costs, which is three times greater than the cost of design and construction [2]. This demonstrates the significance of implementing intelligent asset documentation and structural health monitoring (SHM) approaches for existing built assets. Laser scanning has been widely used to

IntechOpen

document and monitor existing conditions of real-world assets in the form of point clouds [3, 4]. A point cloud is an unstructured low-level digital representation, which by itself does not contain any meaningful information of the documented asset. A 'twinning' process is utilised to convert the low-level data into a high-level digital representation in a structured format, namely, a geometric Digital Twin (gDT) [5]. The gDT can be further enriched with other information, such as semantic meanings, texture, materials, damage, energy use, maintenance data and so forth from its physical twin using IoT technologies [6], to form an information enriched model over time, namely, a 'digital twin' (DT). 'Geometric accuracy' is a vital indicator that guides and describes the degree of spatial accuracy of the resulting twin. It is conventionally deemed as the Represented Accuracy [7] that denotes the standard deviation range to be achieved once the point cloud is twinned into a geometric model. Twinning a real-world asset is an interpretive process, where geometric accuracy largely depends on a modeller's experience and discretion [5]. While in their unstructured state, point clouds contain more geometric details than a resulting gDT created from the point cloud. Therefore, the resulting 'best-fit' gDTs are highly unlikely to be as accurate as the measured data (e.g., a point cloud) at the end of the twinning process [8]. This is also true for the automated methods since there is a trade-off between the achieved geometric accuracy and the quantity of information used for describing existing constructive objects in arbitrary shapes [9]. This occurs because the process of twinning involves simplifications to create polygon- or mesh-based primitives so that it 'smooths' discontinuities and gaps in point clouds [10]. This means that almost every object is approximated in order to transform point-cloud-based descriptors (in non-parametric formats) into parametric primitives [11]. **Figure 1** illustrates a series of components for a bridge asset where the point cloud is converted into bespoke gDT elements. However, since point clouds often contain defects, such as varying point density [12] and occlusions [13], it is difficult or often not feasible to achieve a desired LOGA for resulting gDTs [5]. When these conditions occur, what are realistic expectations for a modeller or of an automated method with regard to representing the reality and meeting the required accuracies for SHM?

Numerous specifications termed as LOX (e.g. Level of Development and Level of Detail) have been developed to guide practitioners and researchers when creating digital models [14]. What do the LOX mean? How to measure whether the specifications were met? What is the best practice approach to reflect when the employer

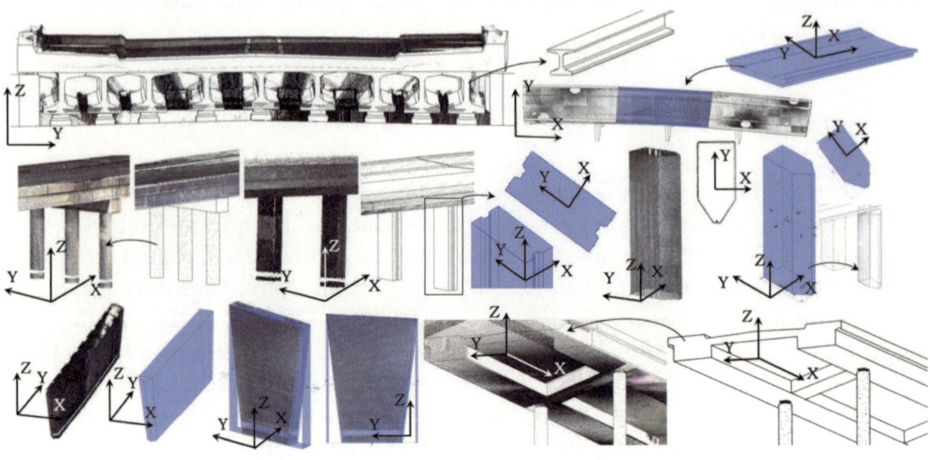

Figure 1.
Customising shapes of bridge components and fitting them to point clusters.

requires '1 cm accuracy' or 'every element to be within a half centimetre'? This chapter explores these questions, aiming (1) to provide a critical review of existing specifications and twinning implementations, (2) to identify technical twinning challenges, and (3) to inform the establishment of a geometric-accuracy-based evaluation system for twinning and updating.

2. Background

2.1 Existing LOX

The term 'LOD' was initially introduced by Vico Software [15]. Ambiguity of defining LOD stems largely from the fact that the American Institute of Architects (AIA) later adopted this concept and kept the acronym LOD but changed it to mean 'Level of Development' rather than 'Level of Detail' [16]. It was then superseded by the document AIA G202™ [17], which defines five progressively detailed levels of completeness: LOD100–LOD500. Based on the AIA protocols, the BIMForum [18] released another LOD specification, which was identical to those published in the AIA's Digital Practice Documents [19], but with two exceptions. First, a new LOD was designated as LOD350. Second, the LOD500 was removed from the specification. The geometric requirements of gDT elements of LOD300, LOD350, and LOD400 are defined in the same way in terms of accuracy. However, this BIMForum document does not elaborate on what is implied by 'accurate' or how to measure it. Bolpagni [20–22] summarised the history of the LOX classification system in **Table 1**. Various new classification systems have been developed to accompany and complement the BIMForum's LOD specification. For example, New Zealand proposed a LOD specification that contains five maturity levels [23], each of which is a sum of different aspects that define the geometry and information of gDT elements. Among these, Level of Detail (LOD) and Level of Accuracy (LOA) do not specify any quantitative standards. Royal Institution of Chartered Surveyors [24] proposed a concept of building survey detail accuracy banding, which defines accuracies to be achieved for different surveyed features when an employer requires a customised geometric accuracy and confidence level. This banding, however, is tailored for designing building settings consisting of cuboids defined by length, width, and height. Similarly, Abualdenien and Borrmann [25] introduced a multi-LOD meta scheme, taking into account the geometric uncertainties by assigning quantitative fuzziness in cm. Again, the usefulness of this scheme in describing the twinning quality is unknown. To this end, Banfi [26] and Banfi et al. [27] proposed a new Grades of Generation (GoG) protocol for twinning highly complex historic structures from point clouds. LOGA was defined as the error resulting between the reconstructed objects and the point clouds using metrics such as the mean distance, median distance, and standard deviation. The USIBD specifications [7] were the first to provide the means to report twinning results of existing building conditions (from point clouds) based on standard deviation (stdev). It articulates the 'accuracy' as well as the five different LOAs (**Figure 2**) by which to represent real-world out-of-plumb geometries. Specifically, the Measured Accuracy represents the stdev range that is to be achieved to acquire a point cloud, regardless of the method used. In contrast, the Represented Accuracy represents the stdev range that is to be achieved when a point cloud is twinned. This guideline, however, does not indicate how to achieve and how to measure the Measured Accuracy and Represented Accuracy. As shown, various acronyms are used across countries and organisations. These acronyms are either identical or interchangeable, making them very challenging to be understood or adopted.

Country/ region	Document	Year	LOX	Whole gDT	gDT element	Geometric data/info	Non-geometric data/info
Denmark	BIPS	2007	Information Level	√	√	√	√
Australia	CRC	2009	Object Data Levels/Level of Detail		√	√	√
USA	Department of VA	2010	Level of Development		√	√	√
USA	Vico Software	2011	Level of Detail	√	√	√	√
Australia	NATSPEC	2011	Level of Development		√	√	√
Hong Kong	HKIBIM	2011	Level of Detail		√	√	√
USA	NYC DDC	2012	Model Level of Development/ Level of Development	√	√	√	√
			Model Granularity		√	√	√
	Penn State University	2012	Level of Development		√	√	√
	USC	2012	Level of Detail		√	√	
	US Army Corps of Engineers (USACE)	2012	Level of Development		√	√	√
Singapore	BCA	2013	Level of Detail		√	√	√
UK	PAS 1192–2	2013	Level of Model Definition		√	√	√
			Level of Model Detail		√	√	
			Level of Model Information		√		√
UK	CIC BIM Protocol	2013	Level of Detail	√	—	—	—
Germany	BMVBS	2013	Level of Development		√	√	√
Netherland	BIM	2014	Information Level	√	√	√	√
Canada	AEC	2014	Level of Development	√	√		
France	Le Moniteur	2014	Level of Detail/ Level of Development		√	√	√
Australia	BCPP	2014	Level of Development		√	√	√
			Level of Detail		√	√	
			Level of Accuracy		√	√	√
			Level of Information		√		√
			Level of Coordination	—	—	—	—

Country/ region	Document	Year	LOX	Whole gDT	gDT element	Geometric data/info	Non-geometric data/info
China	CBC	2014	Level of Detail	√	√	√	√
Belgium	ABEB-VBA	2015	Level of Development		√	√	√
Germany	D&R	2015	Level of Development	√		√	
USA	BIMForum	2015	Level of Development		√	√	√
			Element Geometry		√	√	
			Associated Attribute Information		√		√
UK	NBS BIM Toolkit	2015	Level of Detail		√	√	
			Level of Information		√		√
UK	AEC (UK)	2015	Level of Definition		√	√	√
			Level of Information		√		√
			Grade/Level of Detail		√	√	
China	SZGWS (Shenzhen)	2015	LOD	√		√	√

Table 1.
Comparison of the LOX classification system across countries.

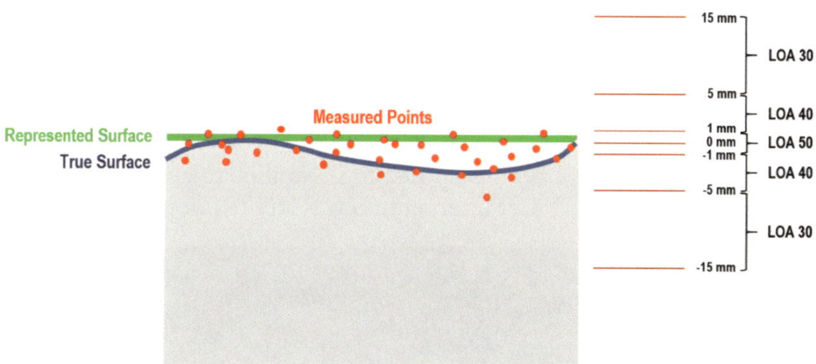

Figure 2.
Measured and represented accuracy [7].

2.2 Industry applications

Leading software vendors provide advanced commercial twinning solutions, which are currently semi-automated processes at best. *ClearEdge3D Edgewise* software can automatically extract geometric features for industrial constructive elements and basic architectural elements using cross-sections in user-cropped regions followed by fitting 3D shapes from a library of preloaded features [28, 29]. This means, the current practice can achieve a high degree of automation of twinning if the resulting geometries are assumed to be generic or pre-defined. However, in the context of SHM, this assumption is unrealistic if a millimetre twinning accuracy is required. Twinning arbitrary geometries using point clouds is quite challenging [30]. Most authoring tools are designed to model orthogonally, or along local coordinate axes. They employ the use of rigid-body parameters to design construction elements by defining cross-sectional shapes, length, width and height parameters, whereas in the real world, as-is components are often warped, off-plumb, or contain deflections [31]. While finite element analysis and multi-physics engines can be used to predict elastic and plastic distortions in materials [32], current digitization workflows that produce parametric objects cannot capture distortion such as bowing in a beam or welding distortion in steel frames. Errors are introduced when the as-is geometries are twinned as being plumb and subjected to rigid-body physics [33]. In this case, geometry deviation analysis is important because unfitted geometries would potentially reduce the reliability of the gDT to be used for structural analysis and defect detection for SHM purposes. Current authoring applications are not capable of carrying out geometry deviation analysis for point clouds. The actual geometry deviation analysis requires third-party middleware software to interpret and investigate. *FARO BuildIT Construction* [34] is the most recent verification software for dimensional quality control (QC) process. Measured data collected from laser scanners can be compared against a gDT to analyse geometric deviations (**Figure 3**). However, it is worth noting that the nature and origin of a deviation is not identified in the analysis directly. Specifically, the analysis itself is often in the form of a 'heat map', where deviations are plotted in colours that correspond to a specific magnitude and direction from a perfect state (i.e. 0 mm deviation). However, point clouds contain voids and sparse measurements, which as directly classified deviations. These false positive measurements make it difficult to interpret the deviation analysis results. Users must manually inspect datasets to observe and detect gross errors or missing components. Currently, there are no available automated solutions for this in existing middleware. In addition, once deviations are identified through deviation analysis and manual interpretation, users must also manually apply changes to update the authoring gDTs. This is currently a large challenge since there is very little research into automated updating of gDT from point clouds [35, 36].

2.3 Existing research methods

Automated methods have been proposed to streamline the twinning process (**Table 2**) [37, 38]. However, user intervention was still required for some crucial

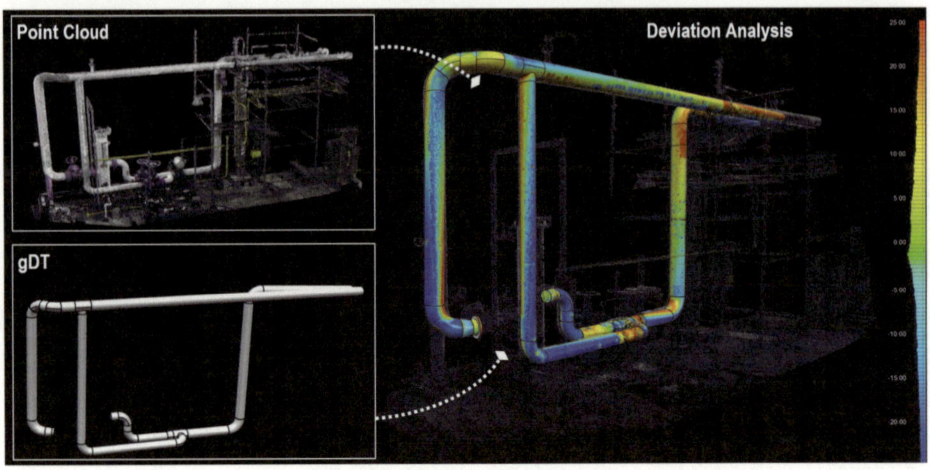

Figure 3.
Deviation analysis of a pipe assembly (point cloud data courtesy of FARO Technologies, Inc.).

	Point cloud authenticity real (R) or synthetic (S)	Manual (×)/automated (√) twining process	Structure completeness	LOGA assessment availability	Manual (×)/ automated (√) LOGA assessment	Micro (I)/macro (A) levels LOGA assessment	LODGA assessment method/metrics
[39]	R	×	×	√	×	A	Minimum Euclidian distance and thresholding
[40]	S	√	√	×	—	—	—
[41]	S	√	√	√	√	A	Hausdorff distance
[42]	R	√	×	√	√	I	Centroid Euclidean distance, area difference (width and length), angular difference
[38]	R	√	√	√	×	A	Visual assessment
[43]	R	√	√	√	√	A/I	Plane fitting/orientation/dimensional error, positioning/sizing error
[44]	R	×	×	√	√	A	Mean point-surface distance
[45]	R	×	√	√	×	A	Progressive densification
[37]	R	√	×	√	×	I	Cylinder radius and orientation
[46]	R	×	√	√	×	A/I	CloudCompare
[47]	R	√	×	×	—	—	—
[11]	R	√	√	√	×/√	A/I	Control points and point-to-gDT
[27]	R	×	×	√	√	A	Standard deviation
[48]	R	×	×	√	√	A	Standard deviation
[49]	R	—	×	√	√	I	Thresholding
[50]	—	×	√	×	—	—	—
[5]	R	√	√	√	√	A	Cloud-to-cloud distance

Table 2.
Twinning methods and evaluation metrics.

steps [44]. Zhang et al. [40] and Laefer and Truong-Hong [47] produced gDTs for bridges and industry plants, but without a geometric deviation assessment. Anil et al. [39] were among the pioneers who discussed in depth the problem of geometric deviation. They suggested using *minimum Euclidean distance* and *thresholding* [49] as metrics to evaluate the fitting quality (CAD model against point clouds). The deviation analysis at macro level (for the whole structure) was performed using a commercial software application (i.e. Polyworks v9). Bonduel et al. [46] suggested assessing the twinning results at both macro and micro levels. They used *CloudCompare* to analyse the deviations between a point cloud and a manually generated building floor gDT. They also discussed the achieved represented accuracy using LOAs provided by USIBD. Then, *Hausdorff distance* was proposed to measure the fitting deviation of a mesh-based building gDT reconstructed from a synthetic point cloud [41]. Thomson and Boehm [42] suggested using *Euclidean distance* and *area difference* based on the width and length, and angular difference to measure the fitting quality of walls. Although these measurements can assess elementwise quality, they are tailored for generic building walls in cuboid shapes. Similarly, Valero et al. [43] assessed fitting deviations of individual furniture objects and walls using *orientation*, *dimension*, *positioning*, and *sizing* metrics, assuming these objects consist of planar surfaces. Lu et al. [5] proposed an automated fitting method to twin bridge components. They gauged the fitting accuracy using *Cloud-to-Cloud* (*C2C*) *distance* metrics—a similar metric used by Shirowzhan et al. [51]. However, the geometric deviation evaluation was performed only at the macro level. NURBS-based methods [27, 44, 45, 48] were employed to reconstruct geometric surfaces for building, industry plant, and historic building elements. Note that the generation of compound pipes requires user intervention to group a set of cylindrical segments followed by automatically fitting surfaces [44]. Likewise, highly complex historic structures require manual surface generation, although extremely high twinning accuracy was reported [27]. *Point-to-surface* distance metrics were used to evaluate the fitting quality [44, 48]. In contrast, Barazzetti [45] used the commercial package Geomagic Studio to evaluate the fitting accuracy of the NURBS curves through a *progressive densification* (i.e. multi-resolution) approach. As shown, there is no fully automatic method to produce geometrically highly accurate twins for existing assets. Also, more comprehensive evaluation metrics need to be established for assessing twinning quality.

3. Case study

Previous sections have discussed that twinning existing assets using point clouds is restricted by current software tools which are limited in their ability to represent out-of-plumb conditions and non-rigid formations. It is also restricted by the limits of the data itself. This section discusses this problem in detail through a case study.

Laser scanning can sample an object's surface as it exists with highly accurate spatial measurements in the form of 3D points. If the documented object is not straight or plumb, the scanner can capture its geometric status. Theoretically, a terrestrial laser scanner such as the FARO Focus 3D X330 [52] has a ranging error of ± 2 mm at 10 m, equating to a systematic measurement error at around of 1σ at 10 m. However, the measured accuracy is affected by many factors, including the standard deviation of the sensor, registration methods, material type being scanned, low temperature, bad weather, and strong sunlight [53]. The overall twinning error E_T can be expressed as a combination of three primary sources of error:

$$E_T = E_{RA} + E_M + E_R, \tag{1}$$

where E_{RA} is the ranging error associated with the laser scanner, E_M is the measured error introduced during scanning and registration, and E_R is the represented error resulting from the process of scan-to-gDT. It is important to specify the error associated with each source independent of each other since they are assumed to be mutually exclusive. In this chapter, we only focus on discussing the represented error E_R, which is independent of the sensor, or parameters of the documented object, or scanning and registration methods. It is related to the manner with which the measured point cloud is being transformed into the outcome, i.e. a gDT and describes the extent the gDT matches the acquired points.

As mentioned earlier, existing authoring software packages are by nature orthographic modelling tools. The challenge with using these software packages becomes how to represent a structure's up-to-date conditions. To complicate matters further, the as-weathered, as-damaged, or as-deviated information of existing assets further increases the representation difficulty. Fitting deviations will be generated and propagated if these conditions are represented in an over-simplified fashion. In addition, sparseness, hidden, or concealed conditions are often encountered in point clouds, making it difficult or impossible to twin constructive objects with certainty. Thus, E_R is the accumulated error from the geometric deviations and the propagation of data uncertainty.

Figure 4 demonstrates current efforts on parametric bridge design [54]. The essential feature for bridges is the horizontal and vertical alignments, which control the parametric relationships and dependencies between assembly systems and all components. The deck cross-sections are then driven by the bridge alignment curves. They are profiles that are used in conjunction with the alignment to derive the overall 3D shape of the bridge deck.

When SHM and retrofit planning is being performed, accurate as-is condition data is required regardless of the availability of the as-designed parametric information. Point clouds can depict the as-is geometries of an asset using thousands of data points. However, maintaining the dimensional accuracy and geometric fidelity of a given bridge point cloud is challenging because the usefulness of topological and geometric constraints is limited to very simple geometric shapes and spatial relationships. As-is geometries do not exhibit a parametric pattern with respect to the initial primitives used to create the as-designed model. **Figure 5** illustrates the non-orthogonal geometries of a real-world bridge point cloud cannot be fitted using

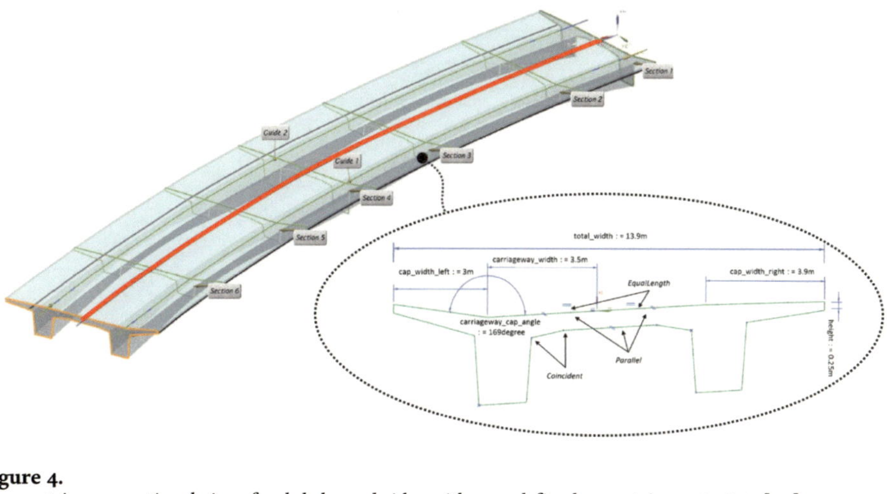

Figure 4.
Parametric cross section design of a slab-beam bridge with user-defined geometric constraints [54].

generic shapes, such as cuboids, in an orthogonal fashion. The modelled slabs do not follow the point cloud and produce fitting deviations when they are joined at sharp angles (**Figure 5a**). These deviations become smaller if the cross-sections are outlined with as-is 2D shapes. However, the bridge gDT does not necessarily close better and become manifold as the fitting quality is improved at the expense of broken or clashing connections (**Figure 5b**). This is especially true when twinning point clouds of pipes with sags, beams and columns with welding distortion or walls that are skewed. Adjacent components do not fit to properly watertight connections unless they are joined at right angles. For example, **Figure 6** illustrates part of a piping system generated using point clouds. The local deviation is reduced from 30 to 1 mm when watertight connections are not used. Given the challenge with the mediation of non-parametric real-world deviations to parametric model primitives, modellers are often forced to leave objects 'slightly off-axis' or perform 'unnatural shape editing' by eliminating or ignoring as many overlapping and joint warnings as possible in order to match the points.

When facing occlusions and damage conditions, the geometric accuracy has a reliance on human perception followed by inferring the hidden information based on assumptions. For example, a bearing plays an important role in a bridge, but its surface is less than 1% of that of the deck slab and has a complex composition. These characteristics make it difficult to be fully captured by a laser sensor (**Figure 7a**). In addition, point clouds need to be down sampled before feeding into in-memory-system-based authoring tools or automated algorithms that cannot handle huge datasets. The down sampling is often performed using a third-party processing

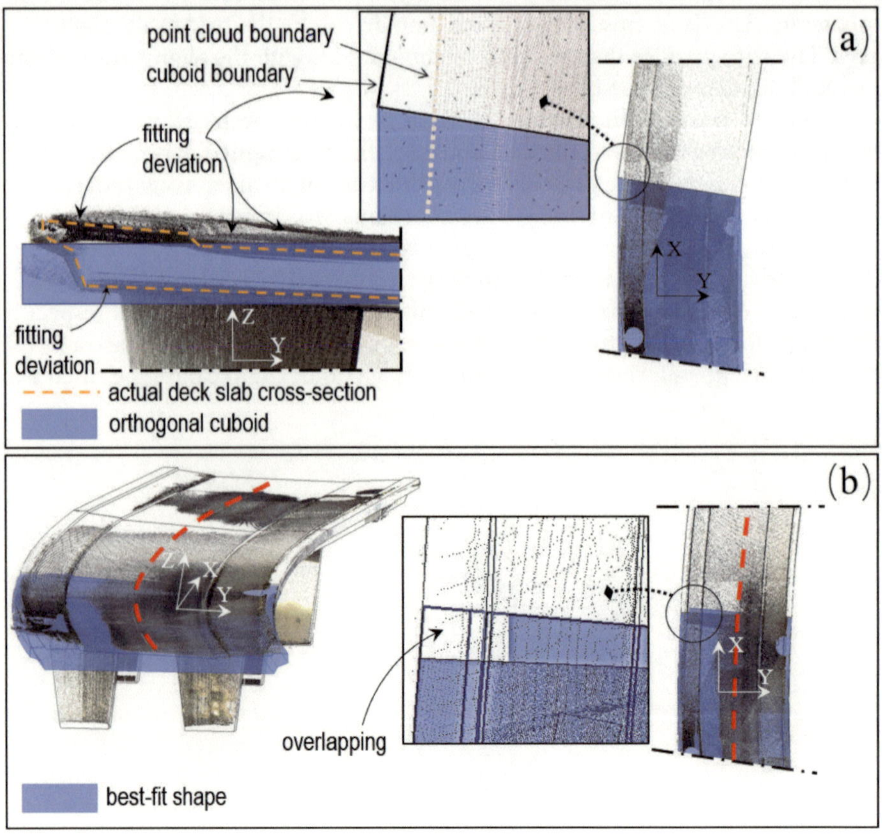

Figure 5.
Fitting geometric shapes to bridge point clouds. (a) Point clouds fitted by cuboids; (b) point clouds fitted by best-fit shapes.

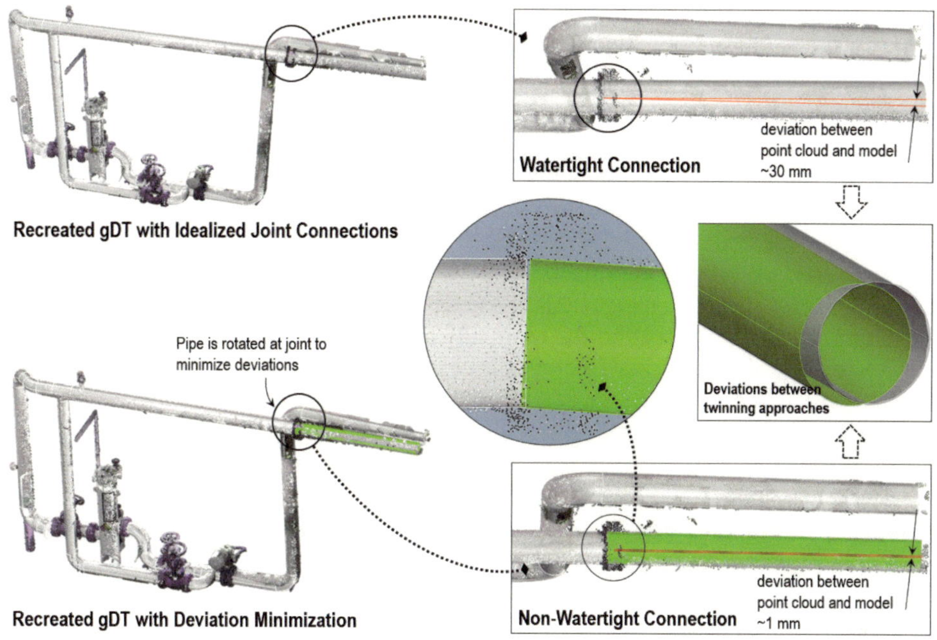

Figure 6.
Fitting cylinders to piping point clouds (point cloud data courtesy of FARO Technologies, Inc.).

software application, which applies generic filters to evenly down sample the points without considering local geometric context. While this is certainly helpful and creates beneficial data compression, the resulting datasets often lose information along the way (i.e. sparse areas or smaller objects will have little to measurements). Thus, only a few points are retained for the bearing surface which does not provide enough information to support the twinning task and result in geometry uncertainties (**Figure 7b**). The interpretation of bearing shapes largely depends on

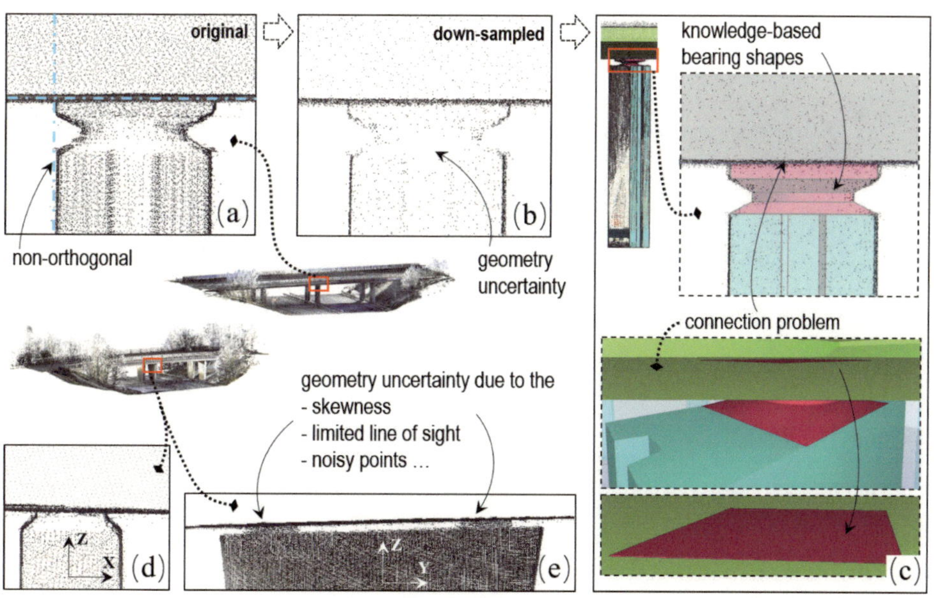

Figure 7.
Bearing gDT generation under uncertainty. (a) Original point cloud; (b) down-sampled point cloud; (c) bearing shapes and connection problem; (d) (e) geometry uncertainty in point clouds.

modeller's knowledge and discretion, which could introduce connection problems (e.g., clashing/gaps) (**Figure 7c**). Uncertainty increases when working with point clouds containing skewness and noise (**Figure 7d** and **e**). Although methods have been suggested to work under occlusions and sparseness [5, 55], the certainty of the resulting models is rarely investigated.

Figure 8 shows an example of a bridge where little-to-no measurements were captured in the girder areas due to a limited line of sight [56]. Like many existing works, both the manual and the automated method inferred specific girder profiles and produced gDTs with detailed dimensions using engineering knowledge. Then, *Cloud-to-Cloud* (*C2C*) distance could be used [5] to compute the deviation between the point clouds sampled from the manually generated gDTs (*Manual*) and the automated ones (*Auto*), and the real point clouds (*Real*):

$$C2C = \max\left(\overline{\text{dist}}_{Manual\ or\ Auto/Real}, \overline{\text{dist}}_{Real/Manual\ or\ Auto}\right), \tag{2}$$

where $\overline{\text{dist}}$ is the estimated distance between a compared point cloud (i.e. Manual or Auto) and a reference point cloud (i.e. *Real*). Non-trivial fitting deviations occurred and raised the overall macro-level deviation ($C2C_{Auto}$—12.5 cm and $C2C_{Manual}$—5.7 cm) [5]. These significant fitting deviations were due to the occluded areas, as no measurements were available to compare against, resulting in an incorrect gDT from a geometric accuracy standpoint. This solution is straightforward since it does not take the modelling uncertainties into account. It simply takes uncertain areas as errors. **Figure 9** illustrates that the fitting deviation was drastically reduced by approximately 70% ($C2C_{Auto}$—4.2 cm) if we replace the complete girder profiles with unclosed mesh-based gDTs while other parts remain

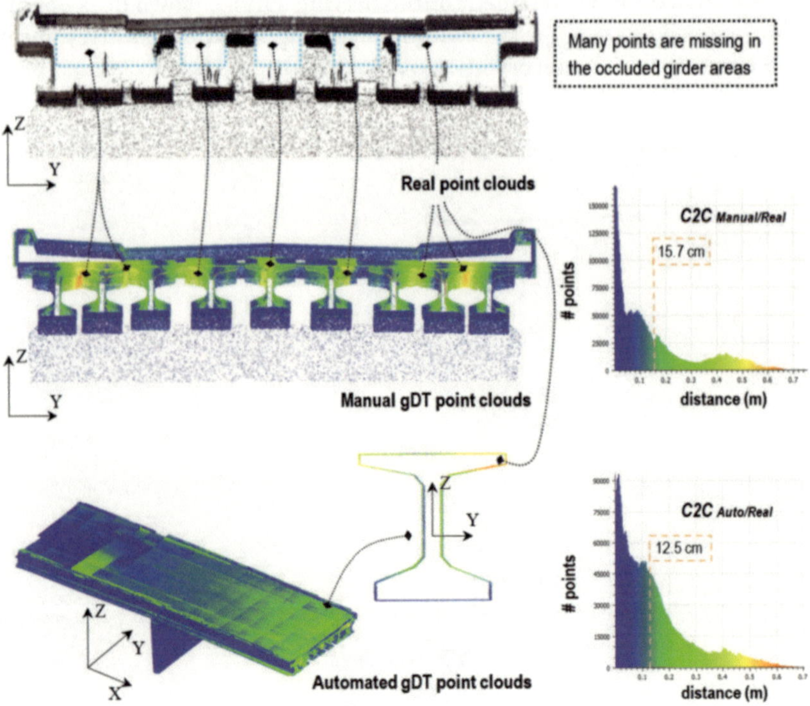

Figure 8.
Geometric deviation with complete girder profiles in occluded areas.

unchanged. Yet still, the improved accuracy only aligns with USIBD's LOA 20 (lower range: 15 mm, and upper range: 5 cm, at 2σ) [7], corresponding to a relatively low accuracy standard. USIBD provides different represented accuracy levels, but it does not specify how to measure it. For example, we can only use a couple of reference points to estimate the accuracy. It is the averaged fraction between pair reference-point distances in the registered scan data and the corresponding pair on-site or gDT point distances:

$$\overline{\text{acc}} = \frac{1}{M}\sum_{i=1}^{M}\left(\frac{\text{pair reference} - \text{point distance}}{\text{pair on} - \text{site or gDT point distance}}\right)_i, \tag{3}$$

where M is the number of investigated pair-wise distance. Then, it is possible to acquire $\overline{\text{acc}}$ that aligns with a higher LOA in USIBD. By contrast, unlike $\overline{\text{acc}}$, *C2C* is an estimation using thousands of calculated points. Therefore, the resulting *C2C*-based accuracy is almost surely not going to achieve an expected 'high accuracy' level (e.g., ±10 mm or USIBD's LOA 30 onwards). The *C2C* comparison between the *Auto* and *Real* revealed that points sampled from bottom flanges of girders were well matched with the original points while the mismatched points were mainly from the central part of the deck slab where points were not evenly distributed. This is attributed to the undulating-surfaces of the gDT generated using the proposed *ConcaveHull alpha-shape* algorithm (**Figure 9**). Local indentations or bumps are generated when *alpha* value is too small to smooth out the surface affected by unavoidable noise, raising the fitting deviations. However, optimising the *alpha* value is difficult because an indentation, for instance, could be due to a defect or a hole but could also due to localised sparse and unevenly distributed points. In addition, although the *ConcaveHull alpha-shape* algorithm can describe slab geometries in a 2D space, it oversimplifies a 3D space.

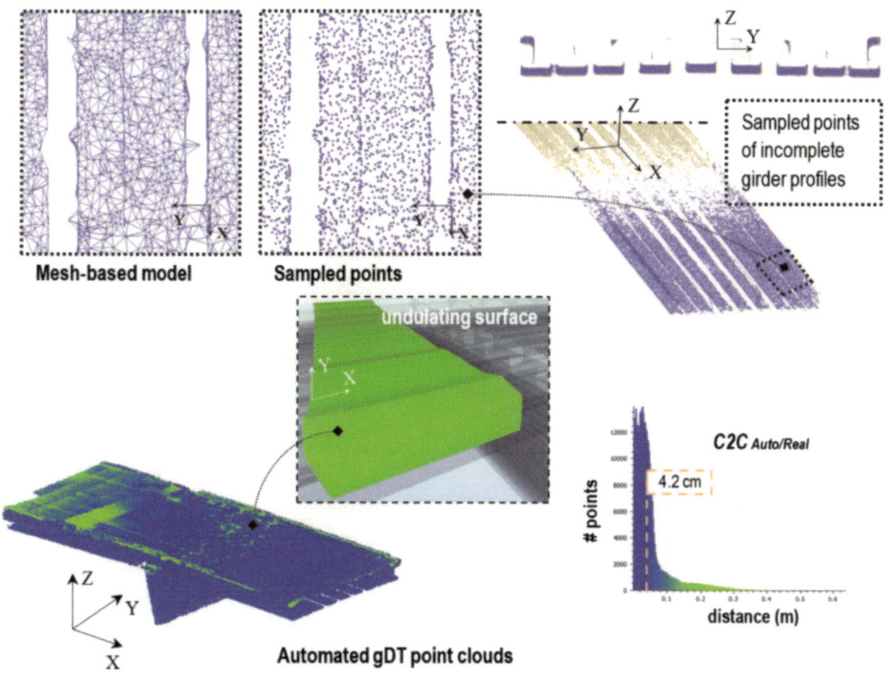

Figure 9.
Geometric deviation with incomplete girder profiles in occluded areas.

4. Prospective twinning methods and deviation analysis

The analysis provided in the previous section demonstrates that real-world conditions are seldom orthogonal and perfect, rendering it extremely difficult to perform high-fidelity twinning with a geometric accuracy on the millimetre scale. Commonly used representation models include but are not limited to: implicit representation such as mathematical formula-based methods [57], Boundary Representation such as polygon- and mesh-based methods [41], Constructive Solid Geometry [58], Swept Solid Representation [47], and NURBS representation [45, 48]. Depending on the nature of defects, the as-damaged geometries may be represented in different ways. **Figure 10** illustrates the vision of the concept of an as-damaged bridge gDT implemented for the inspection work. The method proposed by Hüthwohl et al. [59] can be used to integrate superficial defects such as cracks, efflorescence, corrosion, and slight spalling [**Figure 10a**—(3) and (4)] to the affected element using the back-project technology [59] (**Figure 10b**). In contrast, major defects, such as severe spalling, cavity and pothole [**Figure 10a**—(1) and (2)], are significantly different in geometry compared to their surrounding healthy (i.e. good condition) surfaces. The method proposed by Lu et al. [5] can be used to represent healthy elements; however, it cannot describe the unhealthy areas precisely, due to the extrusion-based twinning nature. Finer representation, such as mesh-based and NURBS-based twinning techniques [50], can be employed to handle the geometry complexity of significant defects in a precise manner (**Figure 10b**). The more variable the defect, the greater the geometric twinning needs to rely on non-parametric representation such as mesh format. One promising solution to produce a gDT that takes the as-damage information into account is to first detect unhealthy areas [60], followed by twinning these unhealthy areas using finer twinning techniques based on their type and size. However, the mesh polygon resolution should not degrade the rendered presentation. This requires an intelligent a priori scheme to resample the point clouds based on the geometric complexity of a sampled surface [61, 62].

Construction elements with different scales may require different twinning techniques. For example, extrusions could be efficient for twinning slab segments; however, they cannot be directly applied to bearings. This means a gDT is highly likely to contain more than one data representation type in order to balance its resolution and the LOGA, which very few works have covered in depth. In addition,

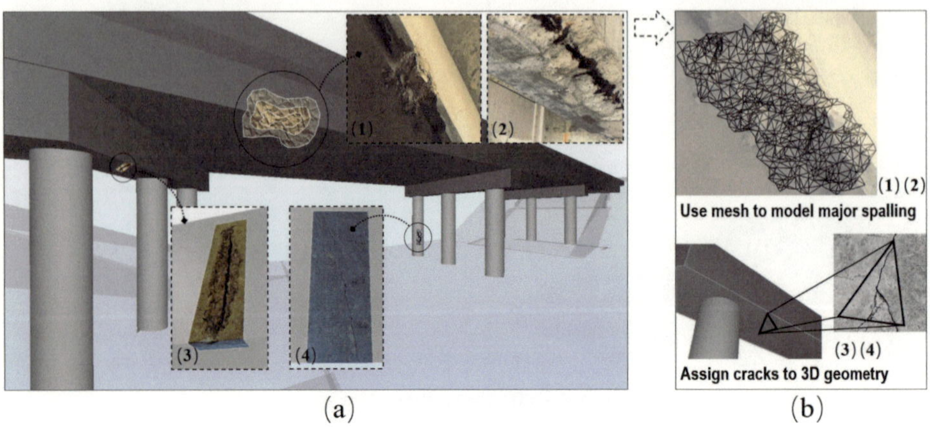

Figure 10.
Vision of the concept of an as-damaged bridge gDT applied for inspection. (a) actual damages or defects; (b) digital representations.

C2C (cm)	Macro [5] (bridge wise)	Micro (element wise)																			
Bridge 1	4.3	Deck slab	Pier cap 1	Pier cap 2	Pier cap 3	Pier 11	Pier 12	Pier 13	Pier 21	Pier 22	Pier 23	Pier 31	Pier 32	Pier 33	—	—	—	—	—	—	—
		6.3	2.8	2.7	2.3	2.2	2.1	1.9	1.8	2.1	2.2	2.2	2.2	1.9	—	—	—	—	—	—	—
	Area	88.0%	2.2% × 3			0.6% × 9									—	—	—	—	—	—	—
Bridge 4	9.4	Deck slab	Pier 11	Pier 12	Pier 13	Pier 14	Pier 15	Pier 16	Pier 21	Pier 22	Pier 23	Pier 24	Pier 25	Pier 26	—	—	—	—	—	—	—
		12.0	3.8	3.6	3.6	3.5	3.7	3.9	3.9	4.3	3.7	3.7	3.9	3.9	—	—	—	—	—	—	—
	Area	89.6%	1%	0.8% × 4				1%	1%	0.8% × 4				1%	—	—	—	—	—	—	—
Bridge 6	4.6	Deck slab	Pier 11	Pier 12	Pier 21	Pier 22	Pier 31	Pier 32	—	—	—	—	—	—	—	—	—	—	—	—	—
		6.9	3.2	3.1	3.4	3.6	2.9	3.3	—	—	—	—	—	—	—	—	—	—	—	—	—
	Area	92.8%	1.2% × 6						—	—	—	—	—	—	—	—	—	—	—	—	—
Bridge 7	12.5	Deck slab	Pier	Girder 11	Girder 12	Girder 13	Girder 14	Girder 15	Girder 16	Girder 17	Girder 18	Girder 19	Girder 21	Girder 22	Girder 23	Girder 24	Girder 25	Girder 26	Girder 27	Girder 28	Girder 29
		6.6	3.0	14	14	13	12	15	17	16	13	13	13	13	14	17	17	12	16	12	13
	Area	54.2%	6.2%	2.1% × 9									2.3% × 9								
Bridge 9	5.6	Deck slab	Pier 11	Pier 12	Pier 21	Pier 22	Pier 31	Pier 32	—	—	—	—	—	—	—	—	—	—	—	—	—
		5.2	3.2	3.0	3.4	3.6	3.9	4.0	—	—	—	—	—	—	—	—	—	—	—	—	—
	Area	84.4%	2.6% × 6						—	—	—	—	—	—	—	—	—	—	—	—	—

Table 3.
Macro- and micro-level C2C geometric deviation analysis.

as previously mentioned, occlusions and sparseness increase the uncertainty of the resulting gDT. These problems require a more intuitive geometric deviation analysis system. The macro-level deviation analysis can provide an overview of the twinning quality whereas it does not reflect a detailed comparison at the component- or feature-level. Therefore, the dimensional QC system of geometric deviation analysis should consist of both macro- and micro-level analysis. The former, can be used to quickly localise uncertain areas, or areas with major deviations (**Figures 8** and **9**) while the latter can provide detailed deviation analysis at the component-level, indicating a more meaningful LOGA of specific elements. **Table 3** shows an example of the *C2C-based* geometric deviation analysis of five bridge gDTs using an automated twinning method. The micro-level numerical indications show that the deck slab takes the bigger part of the overall deviation whereas the other components such as pier caps, piers, and girders take the smaller part. Specifically, for all these bridges except *Bridge 7*, the deviations stemming from deck slabs are 2.9, 3.2, 2.1, and 1.5 times bigger than that of the averaged value for the remaining components, respectively. *Bridge 7* initially appears misleading since the slab deviations are only 48.8% of that of its girders. However, these abnormal deviations are due to significant occlusions in the raw data. The distribution of the deviations is not necessarily proportional to the LOGA. This can be demonstrated through the coverage area of components. The deck slab takes most of the sampled surface compared to that of the pier caps and piers, which are much smaller in size and in covered area. Specifically, pier caps, piers, and girders take 12, 10.4, 7.2, 31.7, and 15.6% of the overall sampled surface of each bridge, respectively. This means although the absolute twinning accuracy of smaller components is higher than larger ones, their relative accuracy is not necessarily better. A deviation analysis system that combines both macro- and micro-level information can better interpret the twinning accuracy.

5. Conclusions

This chapter presents an exploratory analysis of the LOGA of geometric twinning for existing assets using point clouds. Twinning existing assets for monitoring the structural health is a daunting task since the as-is geometric conditions can differ from the designed status due to geometric anomalies, physical damages, deflections, and the complexity, ambiguities, and defects in the measured point cloud data. Section 2.1 reviews existing LOX systems that lack a clear elaboration on geometry accuracy. They share the same acronym but do not necessarily carry the same meaning. They are tailored for basic assumptions made in the design phase or at the beginning of a generative process, making them useless to interpret the as-is geometries of gDTs delivered for SHM purposes. Section 2.2 reviews on industry applications and reveals that there remains a gap between the accuracy requirements placed on gDTs and the capabilities of underlying twinning processes. Specifically, there are practical limitations of authoring tools with respect to the context of orthogonal (i.e. idealised parametric primitives) and real-world deviations (i.e. non-parametric data formats such as point clouds and meshes). Their ability to twin or capture non-rigid-body deformations is extremely limited. Likewise, limitations are also revealed for the deviation evaluation tools with respect to geometric accuracy interpretations. Despite the growing state of the art (Section 2.3), a fully automated twinning and updating process is still in its infancy. A major bottleneck for complete automation of the workflow is the definition of LOGA of the documented asset that covers all geometric deviations and data uncertainties. This requires a development of comprehensive LOGA-based evaluation metrics for gDTs generated in the post-construction stage. The case study (Section 3) demonstrates the technical challenges of the twinning process. High-fidelity twinning within millimetre-level geometric accuracy is challenging to achieve because each step introduces errors. This requires in-depth research on the level of the model certainty. LOGA is closely related to the tools, techniques, and process used to represent the specific object being documented. In the end, the twinning method and LOGA depend highly on what the gDT will be used for (Section 4), on the specific needs and goals of the project, and what kind of metadata is required when providing information about the geometric accuracy.

Parameterising point cloud data results in a loss of geometric accuracy along with a decrease of model certainty. This requires practitioners and researchers to effectively communicate the LOGA through a universal consensus before developing, evaluating, and using gDTs. Until there is a consensus and a universal system for describing geometric accuracy of gDTs, the following recommendations are provided. In the case where geometric accuracy requirements are very strict, such as in the O&M stage, it may be useful to store and link the initial as-is captured data along with the resulting gDT. The purpose for this is two-fold. First, it allows for an end-user to view the initial dataset that was used to create the gDT, for conducting its own unique accuracy or structural analysis. Storing the initial raw point cloud data will provide a level of confidence to an end-user when they use the geometric information from a gDT. It also alleviates some of the burden placed on individuals who create the gDT to provide a subjective global accuracy figure (which can have legal impacts depending on end-use of such gDTs). Secondly, linking the initial data capture avoids loss of geometric data. Since point cloud data contains much rawer geometric information than a resulting surface-based or solid-based gDT, data fidelity can be preserved. As twinning processes and algorithms continue to develop and improve (both in accuracy but also in computational efficiency) it will be possible to build, update, manage, and exploit gDTs in a progressive manner.

Acknowledgements

This research work is supported by the National Sciences and Engineering Research Council (NSERC), Mitacs and Edge Architects Ltd. and Cambridge Trimble Fund. We would like to thank them for their support. We also acknowledge Faro Technologies for their in-kind support, provision of sample point cloud data and access to BuildIT Construction software. Any opinions, findings, and conclusions or recommendations expressed in this work are those of the authors and do not necessarily reflect the views of the stakeholders who have supported this research.

Author details

Ruodan Lu[1,2]*, Chris Rausch[3], Marzia Bolpagni[4], Ioannis Brilakis[5] and Carl T. Haas[3]

1 School of Architecture, Building and Civil Engineering, Loughborough University, United Kingdom

2 Darwin College, University of Cambridge, United Kingdom

3 Department of Civil and Environmental Engineering, University of Waterloo, Canada

4 Mace Group, United Kingdom

5 Department of Engineering, University of Cambridge, United Kingdom

*Address all correspondence to: r.lu@lboro.ac.uk; rl508@cam.ac.uk

References

[1] Shea RH. Historian Uses Lasers to Unlock Mysteries of Gothic Cathedrals. National Geographic Magazine. 2019. Available from: https://news.nationalgeographic.com/2015/06/150622-andrew-tallon-notre-dame-cathedral-laser-scan-art-history-medieval-gothic/ [Accessed: 17 March 2020]

[2] Aziz ND, Nawawi AH, Ariff NRM. Building information modelling (BIM) in facilities management: Opportunities to be considered by facility managers. Procedia - Social and Behavioral Sciences. 31 October 2016;**234**:353-362. DOI: 10.1016/j.sbspro.2016.10.252

[3] Wang Q, Kim MK. Applications of 3D point cloud data in the construction industry: A fifteen-year review from 2004 to 2018. Advanced Engineering Informatics. January 2019;**39**:306-319. DOI: 10.1016/j.aei.2019.02.007

[4] Kim MK, Wang Q, Li H. Non-contact sensing based geometric quality assessment of buildings and civil structures: A review. Automation in Construction. April 2019;**100**:163-179. DOI: 10.1016/j.autcon.2019.01.002

[5] Lu R, Brilakis I. Digital twinning of existing reinforced concrete bridges from labelled point clusters. Automation in Construction. September 2019;**105**: 102837. DOI: 10.1016/j.autcon.2019.102837

[6] Davila Delgado JM, Butler LJ, Gibbons N, Brilakis I, Elshafie MZEB, Middleton C. Management of structural monitoring data of bridges using BIM. Proceedings of the Institution of Civil Engineers - Bridge Engineering. September 2017;**170**(3):204-218. DOI: 10.1680/jbren.16.00013. ISSN 1478-4637. E-ISSN 1751-7664

[7] USIBD. Level of Accuracy (LOA) Specification Version 2.0. U.S. Inst. Build. Doc. 2016. Available from: https://usibd.org/product/level-of-accuracy-loa-specification-version-2-0/ [Accessed: 17 March 2020]

[8] Qu T, Sun W. Usage of 3D point cloud data in BIM (building information modelling): Current applications and challenges. Journal of Civil Engineering and Architecture. 2015;**9**(11):1269-1278

[9] Weisstein EW. Torus. MathWorld—A Wolfram Web Resource. 2018. Available from: http://mathworld.wolfram.com/Torus.html [Accessed: 17 March 2020]

[10] Lu R. Automated generation of geometric digital twins of existing reinforced concrete bridges [Dr. thesis]. 2019. DOI: 10.17863/CAM.36680

[11] Macher H, Landes T, Grussenmeyer P. From point clouds to building information models: 3D semi-automatic reconstruction of indoors of existing buildings. Applied Sciences. 2017;**7**(10):1030

[12] Truong-Hong L, Laefer DF. Quantitative evaluation strategies for urban 3D model generation from remote sensing data. Computers and Graphics. 2015;**49**:82-91

[13] Dimitrov A, Golparvar-Fard M. Segmentation of building point cloud models including detailed architectural/structural features and MEP systems. Automation in Construction. 2015;**51**(C):32-45

[14] Succar B. Level of X (LoX), BIM Dictionary. Available from: https://bimdictionary.com/en/level-of-x/1/ [Accessed: 17 March 2020]

[15] Trimble. VICO Software and Trimble to Integrate Workflows and Use Building Information Models to Construct Better Buildings. 2008. Available from: http://investor.trimble.

com/news-releases/news-release-details/vico-software-and-trimble-integrate-workflows-and-use-building

[16] AIA. Building Information Modeling Protocol Exhibit. AIA Doc. E202TM; 2008. Available from: https://www.smacna.org/docs/default-source/building-information-modeling/aia-e202-building-information-modeling-protocol-exhibit-pdf.pdf?sfvrsn=333afea5_0 [Accessed: 03 June 2020]

[17] AIA. AIA Document E203—2013 Building Information Modeling and Digital Data Exhibit; 2013. Available from: http://content.aia.org/sites/default/files/2016-09/AIA-E203-2013-Free-Sample-Preview.pdf [Accessed: 03 June 2020]

[18] Reinhardt J et al. Level of Development Specification for Building Information Models. 2013. Available from: https://bimforum.org/resources/Documents/BIMForum_LOD_2013_reprint.pdf [Accessed: 03 June 2020]

[19] AIA. Digital Practice Documents—Guide, Instructions and Commentary to the 2013 AIA Digital Practice Documents. AIA Document E202TM; 2013. Available from: https://help.aiacontracts.org/public/wp-content/uploads/2020/03/Digital-Practice_Guide.pdf [Accessed: 03 June 2020]

[20] Bolpagni M, Ciribini ALC. The information modeling and the progression of data-driven projects. In: Proceedings of the CIB World Building Congress 2016. Vol. 3. Building up Business Operations and Their Logic. Shaping Materials and Technologies; 2016

[21] Bolpagni M. The Implementation of BIM within the Public Procurement: A Model-Based Approach for the Construction Industry. 2013. Available from: https://www.vttresearch.com/sites/default/files/pdf/technology/2013/T130.pdf [Accessed: 03 June 2020]

[22] Bolpagni M. Digitalisation of tendering and awarding processes: A Building Information Modelling (BIM)-based approach to public procurement routes [Doctoral dissertation]. Politecnico di Milano; 2018. Available from: http://hdl.handle.net/10589/142140

[23] Reding A, Williams J. Appendix C—Levels of Development Definitions. New Zealand BIM Handbook. 2014. Available from: https://www.building.govt.nz/assets/Uploads/projects-and-consents/building-information-modelling/nz-bim-handbook-appendix-c-levels-of-development-definitions.pdf

[24] RICS. Measured Surveys of Land, Buildings and Utilities. 3rd ed. RICS Prof. Guid. Glob. Royal Institution of Chartered Surveyors (RICS); 2014. Available from: https://www.rics.org/globalassets/rics-website/media/upholding-professional-standards/sector-standards/land/measured-surveys-of-land-buildings-and-utilities-3rd-edition-rics.pdf

[25] Abualdenien J, Borrmann A. Multi-LOD model for describing uncertainty and checking requirements in different design stages. In: eWork and eBusiness in Architecture, Engineering and Construction. 2019. DOI: 10.1201/9780429506215-24

[26] Banfi F. BIM orientation: Grades of generation and information for different type of analysis and management process. International Archives of the Photogrammetry, Remote Sensing and Spatial Information Sciences—ISPRS Archives. 2017;**XLII-2/W5**:57-64. DOI: 10.5194/isprs-archives-XLII-2-W5-57-2017

[27] Banfi F, Fai S, Brumana R. BIM automation: Advanced modeling generative process for complex structures. In: ISPRS Annals of the Photogrammetry, Remote Sensing and

Spatial Information Sciences. 2017;**IV-2/W2**:9-16. DOI: 10.5194/isprs-annals-IV-2-W2-9-2017

[28] ClearEdge3D. Structure Modelling Tools. 2019. Available from: https://www.clearedge3d.com/ [Accessed: 17 March 2020]

[29] ClearEdge3D. EdgeWise. 2019. Available from: https://www.clearedge3d.com/products/edgewise/ [Accessed: 17 March 2020]

[30] Wang C, Cho YK, Kim C. Automatic BIM component extraction from point clouds of existing buildings for sustainability applications. Automation in Construction. 2015;**56**: 1-13

[31] Russo JM. What is meant by 'level of accuracy?'. LIDAR Magazine. 2014;**4**(3). Available from: http://lidarmag.com/wp-content/uploads/PDF/LiDARNewsMagazine_Russo-WhatIsLevelOfAccuracy_Vol4No3.pdf [Accessed: 17 March 2020]

[32] Castellazzi G, D'Altri AM, Bitelli G, Selvaggi I, Lambertini A. From laser scanning to finite element analysis of complex buildings by using a semi-automatic procedure. Sensors (Switzerland). 2015;**15**(8):18360-18380. DOI: 10.3390/s150818360

[33] Rausch C, Nahangi M, Perreault M, Haas CT, West J. Optimum assembly planning for modular construction components. Journal of Computing in Civil Engineering. January 2017;**31**(1). DOI: 10.1061/(ASCE)CP.1943-5487.0000605

[34] Faro BuildIT Construction. 2019. Available from: https://www.faro.com/products/construction-bim-cim/faro-buildit-construction/ [Accessed: 17 March 2020]

[35] Lin YC, Lin CP, Hu HT, Su YC. Developing final as-built BIM model management system for owners during project closeout: A case study. Advanced Engineering Informatics. April 2018;**36**: 178-193. DOI: 10.1016/j.aei.2018.04.001

[36] Hamledari H, Rezazadeh Azar E, McCabe B. IFC-based development of as-built and as-is BIMs using construction and facility inspection data: Site-to-BIM data transfer automation. Journal of Computing in Civil Engineering. March 2018;**32**(2). DOI: 10.1061/(ASCE)CP.1943-5487.0000727

[37] Patil AK, Holi P, Lee SK, Chai YH. An adaptive approach for the reconstruction and modeling of as-built 3D pipelines from point clouds. Automation in Construction. 2017;**75**:65-78

[38] Ochmann S, Vock R, Wessel R, Klein R. Automatic reconstruction of parametric building models from indoor point clouds. Computers and Graphics. 2016;**54**:94-103

[39] Anil EB, Tang P, Akinci B, Huber D. Deviation analysis method for the assessment of the quality of the as-is building information models generated from point cloud data. Automation in Construction. 2013;**35**:507-516

[40] Zhang G, Vela PA, Brilakis I. Automatic generation of as-built geometric civil infrastructure models from point cloud data. In: Computing in Civil and Building Engineering. Vol. 2014. 2014. pp. 406-413

[41] Oesau S, Lafarge F, Alliez P. Indoor scene reconstruction using feature sensitive primitive extraction and graph-cut. ISPRS Journal of Photogrammetry and Remote Sensing. 2014;**90**:68-82

[42] Thomson C, Boehm J. Automatic geometry generation from point clouds for BIM. Remote Sensing. 2015:7(9): 11753-11775. DOI: 10.3390/rs70911753

[43] Valero E, Adán A, Bosché F. Semantic 3D reconstruction of furnished interiors using laser scanning and RFID technology. Journal of Computing in Civil Engineering. 2016;**30**(4):04015053

[44] Dimitrov A, Gu R, Golparvar-Fard M. Non-uniform B-spline surface fitting from unordered 3D point clouds for as-built modeling. Computer-Aided Civil and Infrastructure Engineering. 2016; **31**(7):483-498

[45] Barazzetti L. Parametric as-built model generation of complex shapes from point clouds. Advanced Engineering Informatics. August 2016; **30**(3):298-311. DOI: 10.1016/j.aei.2016.03.005

[46] Bonduel M, Bassier M, Vergauwen M, Pauwels P, Klein R. Scan-to-BIM output validation: Towards a standardized geometric quality assessment of building information models based on point clouds. In: International Archives of the Photogrammetry, Remote Sensing and Spatial Information Sciences—ISPRS Archives. 2017

[47] Laefer DF, Truong-Hong L. Toward automatic generation of 3D steel structures for building information modelling. Automation in Construction. 2017;**74**:66-77

[48] Brumana R et al. Generative HBIM modelling to embody complexity (LOD, LOG, LOA, LOI): Surveying, preservation, site intervention—The Basilica di Collemaggio (L'Aquila). Applied Geomatics. 2018;**10**:545-567. DOI: 10.1007/s12518-018-0233-3

[49] Stojanovic V, Richter R, Döllner J, Trapp M. Comparative visualization of BIM geometry and corresponding point clouds. International Journal of Sustainable Development and Planning. 2018;**13**(1):12-23. DOI: 10.2495/SDP-V13-N1-12-23

[50] Ma L, Sacks R, Kattel U, Bloch T. 3D object classification using geometric features and pairwise relationships. Computer-Aided Civil and Infrastructure Engineering. 2018;**33**(2):152-164

[51] Shirowzhan S, Sepasgozar SME, Li H, Trinder J, Tang P. Comparative analysis of machine learning and point-based algorithms for detecting 3D changes in buildings over time using bi-temporal lidar data. Automation in Construction. September 2019;**105**: 102841. DOI: 10.1016/j.autcon.2019.102841

[52] FARO. FARO® Laser Scanner Focus3D X 330 Manual. 2015. Available from: https://faro.app.box.com/s/4f908b59hcjjj8mezdr58z6n4qy5neli [Accessed: 17 March 2020]

[53] Lichti DD, Harvey BR. The effects of reflecting surface material properties on time-of-flight laser scanner measurements. In: Geospatial Theory, Processing and Applications. 2002. Available from: http://citeseerx.ist.psu.edu/viewdoc/download?doi=10.1.1.123.5103&rep=rep1&type=pdf. Corpus ID: 9926219 [Accessed: 03 June 2020]

[54] Ji Y, Borrmann A, Beetz J, Obergrießer M. Exchange of parametric bridge models using a neutral data format. Journal of Computing in Civil Engineering. 2013;**27**(6):593-606

[55] Adan A, Huber D. 3D reconstruction of interior wall surfaces under occlusion and clutter. In: Proceedings of the 2011 International Conference on 3D Imaging, Modeling, Processing, Visualization and Transmission (3DIMPVT). 2011. pp. 275-281

[56] Lu R, Brilakis I, Middleton CR. Detection of structural components in point clouds of existing RC bridges. Computer-Aided Civil and Infrastructure Engineering. March 2019; **34**(3):191-212. DOI: 10.1111/mice.12407

[57] Weisstein EW. Ellipsoid. MathWorld—A Wolfram Web Resource. 2018. Available from: http://mathworld.wolfram.com/Ellipsoid.html [Accessed: 17 March 2020]

[58] Deng Y, Cheng JCP, Anumba C. Mapping between BIM and 3D GIS in different levels of detail using schema mediation and instance comparison. Automation in Construction. 2016;**67**: 1-21

[59] Hüthwohl P, Brilakis I, Borrmann A, Sacks R. Integrating RC bridge defect information into BIM models. Journal of Computing in Civil Engineering. 2018; **32**(3):04018013

[60] Xu Z, Li S, Li H, Li Q. Modeling and problem solving of building defects using point clouds and enhanced case-based reasoning. Automation in Construction. December 2018;**96**:40-54. DOI: 10.1016/j.autcon.2018.09.003

[61] Chen J, Zhang C, Tang P. Geometry-based optimized point cloud compression methodology for construction and infrastructure management. In: Computing in Civil Engineering 2017: Smart Safety, Sustainability and Resilience - Selected Papers from the ASCE International Workshop on Computing in Civil Engineering. American Society of Civil Engineers (ASCE). 2017. pp. 377-385

[62] Zhang C, Tang P. Visual complexity analysis of sparse imageries for automatic laser scan planning in dynamic environments. In: Computing in Civil Engineering. 2015. pp. 271-279

Chapter 8

Safety Evaluation of Stay Cables of Cable-Stayed and Extradosed Bridges via Deterministic and Non-deterministic Methods

Khawaja Ali and Aleena Saleem

Abstract

Cable-stayed and extradosed bridges are thought to be identical structures because both bridges use stay cables for reinforcement. However, the safety factors of their stay cables are stipulated differently in many international standards, i.e., Japanese specifications suggest the safety factors of 2.5 and 1.67 for the design of cable-stayed and extradosed bridges, respectively. In this chapter, a parametric study is carried out for the evaluation of safety factors of stay cables by employing the deterministic and nondeterministic methods at limit states. As a result, it is found that the safety factors in the range of 2.3–2.5 and 1.67 are indispensable for the safe design of cable-stayed and extradosed bridges, respectively, to satisfy the conditions of limit states and target reliability index.

Keywords: cable-stayed bridge, extradosed bridge, stay cable, safety factor, reliability, fatigue, limit state

1. Introduction

The extradosed bridge is thought to be a special form of cable-stayed bridge because both bridges use inclined stay cables for supporting the girder load elastically at points along its length in order to increase the span of girder without intermediate piers [1]. The dead and live loads on girders are transferred to towers by axial action of stay cables. Thus, the safety of these kinds of flexible structures is mainly dependent on the safety of stay cables, which is usually assured by introducing a safety factor to provide a margin between theoretical strengths (R) and load effects (S). For instance, the allowable stress (σ_{all}) at serviceability limit state (SLS) as per the Japan Prestressed Concrete Engineering Association's Specifications, may be determined as 0.4 σ_{UTS} and 0.6 σ_{UTS} (where σ_{UTS} is ultimate tensile strength) for the design of cable-stayed and extradosed bridges, respectively [2]. In that context, Ali et al. [3] estimated an optimum value of safety factor for stay cables of a cable-stayed bridge under ultimate and fatigue limit states by considering the effects of various unexpected events. However, the problem of how much σ_{all} should be used for the stay cables of extradosed bridges is still controversial because these cables are considered as external cables arranged outside the box girder. Moreover, the safety factors of these cables have not been verified against extreme loading and unexpected damage conditions. Besides this, the stress range in a stay cable due to live load is one of the most important considerations for the design of stay cables against fatigue failure [4]. Owing to the variations in live loads, it is difficult to precisely examine the safety of these kinds of flexible structures through an evaluation method comprising safety factors based on experience. Therefore, it seems to be reasonable to conduct safety and reliability assessment using a nondeterministic reliability method which takes into account the effects of all kinds of uncertainties [5–7].

In this paper, a parametric study is carried out to evaluate the safety factors of stay cables of cable-stayed and extradosed bridges by employing the deterministic and nondeterministic methods at limit states. The effects of various parameters, i.e., cable loss and deterioration of cables due to corrosion, on demand to capacity ratio (DCR) of stay cables are also considered in this study. Finally, it is found that the safety factors in the range of 2.3–2.5 and 1.67 are essential for the safe design of cable-stayed and extradosed bridges, respectively to satisfy the conditions of limit states and target reliability index.

2. Finite element modeling

2.1 FE model of cable-stayed bridge (CSB)

A 3D FE model of a cable-stayed bridge, with a main span length of 460 m, is developed using a FEM software (Midas Civil). The structural configuration of the bridge model is shown in **Figure 1**. The bridge model is cambered linearly by 2%. The steel box girder is used for this model. The total width and depth of girder are 21.75 m and 3.5 m, respectively with four design lanes of each 3.5 m wide as shown in **Figure 2**. The configuration of tower is an H-shape composed of steel legs. The total height of tower is 140 m and pylon height (110 m) is taken as 1/4th of the main span length. Moreover, cable-stayed bridge model consists of 144 stay cables (Cs), arranged in a modified-fan style. The anchorage points of stay cables at the bridge deck are located at an interval of 12 m. Tower and girder are modeled as elastic beam elements (168 beams) whereas stay cables are modeled as truss elements (only tension). Fishbone modeling technique is adopted to connect the stay cables with deck spine through rigid links. Moreover, the model is supported by roller supports provided on each end of bridge and piers are assumed to be fixed into firm foundation. All bearings of main girder are movable in longitudinal direction of bridge, i.e., there is no connection between tower and girder at their intersection. The attachments of the cables to tower are pinned. Elastomeric rubber bearings are installed to connect the girder with lower transverse beam through elastic links.

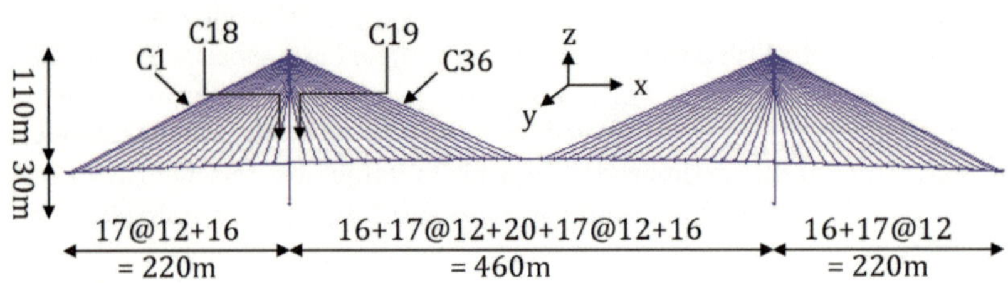

Figure 1.
Configuration of cable-stayed bridge model.

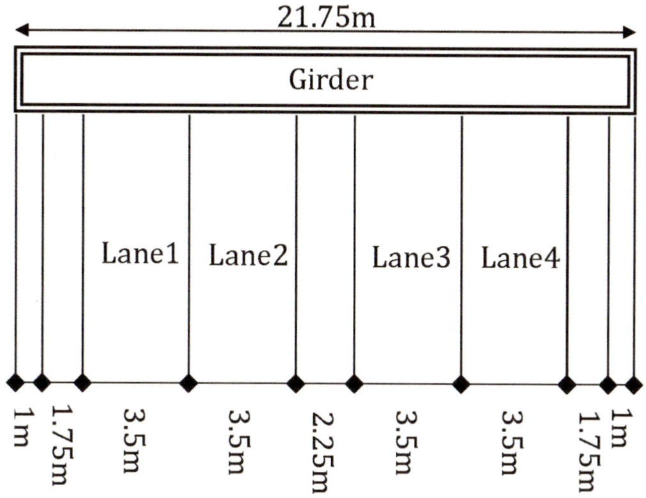

Figure 2.
Configuration of traffic lanes.

2.2 FE model of extradosed bridge (EDB)

Similar to cable-stayed bridge, a 3D FE model of extradosed bridge, with a main span length of 208 m and two side spans of each 100 m, is developed. The structural configuration of bridge model is shown in **Figure 3**. The total width and depth of concrete bridge girder are 21.75 m and 4.5 m, respectively with four lanes as already shown in **Figure 2**. The depth of girder is kept same at the pylon locations as well as at mid-span. The total height of the concrete tower is 40 m and pylon height (20 m) is taken as 1/10th of the main span length. The bridge girder is supported by the piers and a system of 88 stay cables (EDCs) arranged in a modified-fan style. The anchorage points of stay cables (EDCs) at the bridge deck are located at the intervals of 5 m and 6 m on side and main spans, respectively. The connection between tower and girder is assumed to be fixed and monolithic because stress range due to live load in the cables is affected by the girder stiffness and fixity of support on the piers. When the girder is stiff, the stress range in cables due to live load will be small in comparison with permanent loads. To reduce the magnitude of this stress range, girder should be fixed at the piers.

2.3 Design considerations for cable-stayed and extradosed bridges

Bridge design loads are referred to Japanese specifications for highway bridges [4] as shown in **Table 1**. Dead loads are applied uniformly on entire spans whereas

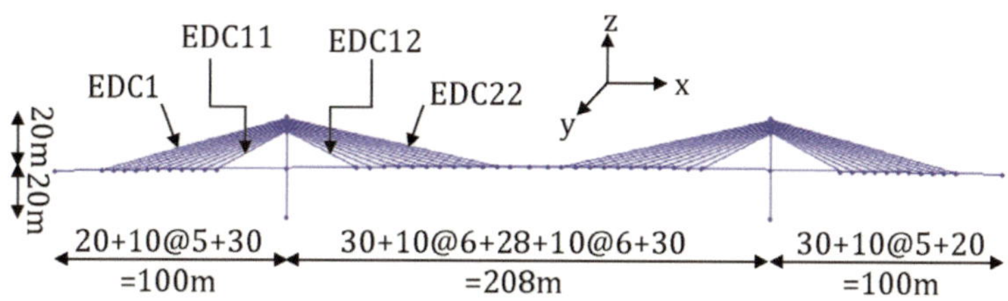

Figure 3.
Configuration of extradosed bridge model.

Dead load, DL_{CSB} (kN/m)	Self-weight of deck	w_{DC1}	48.5
	Pavement loads	w_{DW}	34.2
	Additional loads	w_{DC2}	4.85
Dead load, DL_{EDB} (kN/m)	Girder self-weight	w_{DC1}	335
	Pavement loads	w_{DW}	34.2
	Additional loads	w_{DC2}	4.85
Live load, LL (kN/m)	Concentrated load	P_1	97.5
	Uniformly dist. Load	P_2	29.3
	Pedestrian load	PL	10

Table 1.
Design loads.

B-live loads (concentrated live load: P_1 and uniformly distributed load: P_2) are applied only on main spans of both bridges. The material and sectional properties of bridge components are also shown in **Tables 2** and 3, respectively.

2.4 Stay cables of cable-stayed bridge (Cs)

Preliminary design of stay cables of cable-stayed bridge (Cs) is carried out by assuming a safety factor of 2.5 against σ_{UTS} following the allowable stress design

Properties	Stay cables of CSB	Stay cables of EDB
σ_{UTS} (MPa)	1860	2000
σ_y (MPa)	1302	1400
σ_{all} (MPa)	744	1200
E (GPa)	195	195
ν	0.3	0.3
γ (kN/m^3)	77	77

Table 2.
Material properties of stay cables.

	Members	Deck	Pylon	Pier	Transverse beam
CSB	A (m^2)	0.59	1.11	1.11	0.55
	I_{xx} (m^4)	14.73	7.96	7.96	2.61
	I_{yy} (m^4)	5.13	6.24	6.24	2.14
	I_{zz} (m^4)	29.03	4.72	4.72	1.52
EDB	A (m^2)	13.54	6	12	6
	I_{xx} (m^4)	168.62	4.7	19.44	4.7
	I_{yy} (m^4)	54.22	4.5	16	4.5
	I_{zz} (m^4)	683.84	2	9	2

Table 3.
Sectional properties of bridge components.

(ASD) method. An optimization technique of finding unknown load factors is applied to find the initial pretension forces (PS) through an iterative process in order to achieve the balanced state of bridge under its own weight. Subsequently, the cross-sectional areas of stay cables are calculated and shown in **Figure 4**. In addition to that, stay cables are designed in such a way that axial stresses in stay cables are about 50–60% of σ_{all} under dead loads and less than 95% of σ_{all} under dead plus live loads.

2.5 Stay cables of extradosed bridge (EDCs)

Similar to cable-stayed bridge, the preliminary design of stay cables of extradosed bridge (EDCs) is also carried out by using a safety factor of 1.67. For the calculation of initial pretension forces (PS) of stay cables, the continuous beam method is applied. Hit and trial method is used to find the ideal and balanced state of extradosed bridge under dead loads. Many iterations are performed to optimize the bending moment and cable forces, and cross-sectional areas of stay cables are calculated accordingly as shown in **Figure 5**. In extradosed bridge, the prestress force (Pi) is also applied to the concrete girder. Full pre-stressing of the girder is not feasible. Since only concentric pre-stressing can be used locally in the girder (eccentric pre-stressing causes a secondary bending moment as large as the primary bending moment), a prestress force (Pi) of 200,000 kN is required at main span and some portion of side span to keep the girder un-cracked. Pi is required to

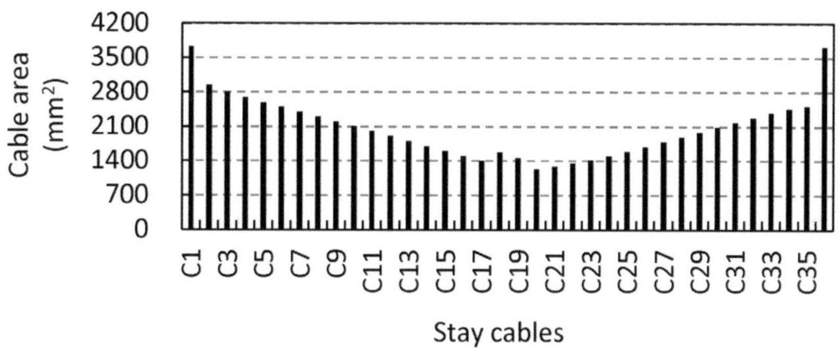

Figure 4.
Cross-sectional areas of stay cables of cable-stayed bridge.

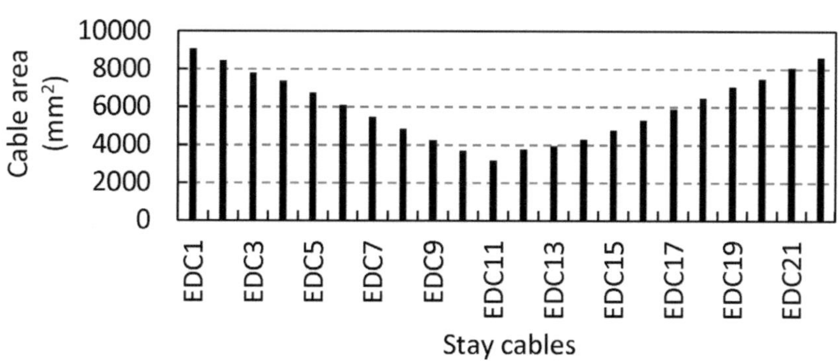

Figure 5.
Cross-sectional areas of stay cables of extradosed bridge.

minimize the deflection and to resist the bending moments due to long-term effects and live loads.

2.6 Effects of nonlinearity

Nonlinearity effects including cable sag effect due to self-weight of stay cables and P-Delta effects due to interaction of deck and tower are also considered in the analysis of both bridge types. Reduced or equivalent modulus of elasticity of stay cables is determined by:

$$E_{eq} = \frac{E}{1 + \frac{(wL)^2 AE}{12T^3}} \tag{1}$$

Eq. (1) is known as Ernst' formula in which E_{eq} is equivalent modulus of elasticity, E is effective material modulus of elasticity, A is cross-sectional area of stay cable, w is cable weight per unit length, L is horizontal projected length and T is tensile force in stay cable.

3. Safety evaluation of stay cables by deterministic method

3.1 Fatigue limit state

For the evaluation of safety factor of stay cables at fatigue limit state, moving load analysis is performed by applying fatigue design load (T-load: 200 kN) to the cable-stayed and extradosed bridge models. Then, influence line diagrams (ILDs) of axial forces in stay cables are drawn by using Breslau Muller Principle and maximum and minimum design variables are calculated. **Figure 6** shows the ILDs of axial forces of stay cables (C1 and EDC1) of cable-stayed and extradosed bridges, respectively. It is observed that the area under ILD of C1 is larger than that of EDC1 under the same fatigue load which indicates that extradosed bridge is less influenced by fatigue load as compared to cable-stayed bridge. Subsequently, cable reversal stresses and design stress range ($\Delta\sigma_d$) values are determined by considering the cyclic loads of constant amplitude and fully reversed nature as per the guidelines of fatigue design recommendations for steel structures [8]. To assess the safety factor at fatigue limit state based on equivalent stress range theory, following equation should be satisfied [9]:

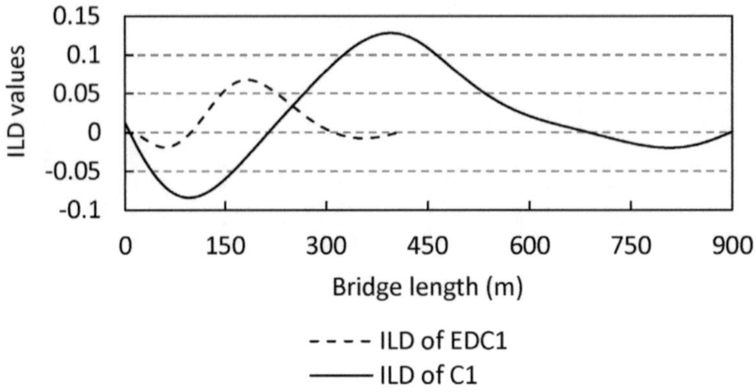

Figure 6.
ILDs of axial forces of stay cables C1 and EDC1.

$$\gamma\left(\frac{\Delta\sigma_d}{\Delta\sigma_R}\right) \leq 1.0 \quad (2)$$

where γ is safety factor equal to 1.2 based on redundancy and importance of structure, $\Delta\sigma_d$ is design stress range also known as maximum stress range and $\Delta\sigma_R$ is allowable stress range which can be found by using Eq. (3):

$$\Delta\sigma_R = \Delta\sigma_{CE} \times C_R \quad (3)$$

where $\Delta\sigma_{CE}$ is the basic allowable stress range or cut off limit for constant amplitude stress which is taken as 270 MPa and 200 MPa for parallel wire strand type stay cables of cable-stayed and extradosed bridges, respectively at 2 million load cycles based on the standard SN or Wohler's curves of cables and C_R is correction factor for mean stress which can be calculated as:

$$C_R = 1.3\left(\frac{1-R}{1.6-R}\right) \text{ for } R \leq -1 \quad (4)$$

$$C_R = \frac{1-R}{1-0.9R} \text{ for } R > -1 \quad (5)$$

which R is the stress ratio defined as the ratio of minimum stress (σ_{min}) to maximum stress (σ_{max}) in stay cables.

Figures 7 and **8** compare the fatigue stress demand to capacity ratios (DCRs) of stay cables of cable-stayed and extradosed bridges, respectively. In case of cable-stayed bridge, stay cable C15 shows maximum DCR under fatigue design load and there is a hefty variation in DCR of stay cables depending on their locations with respect to tower-deck intersection. From **Figure 7**, it can be concluded that a minimum safety factor of 2.2 is necessary to satisfy the fatigue limit state.

In case of extradosed bridge, all stay cables (EDCs) exhibit almost same DCR irrespective of their locations with respect to tower-deck intersection. **Figure 8** also

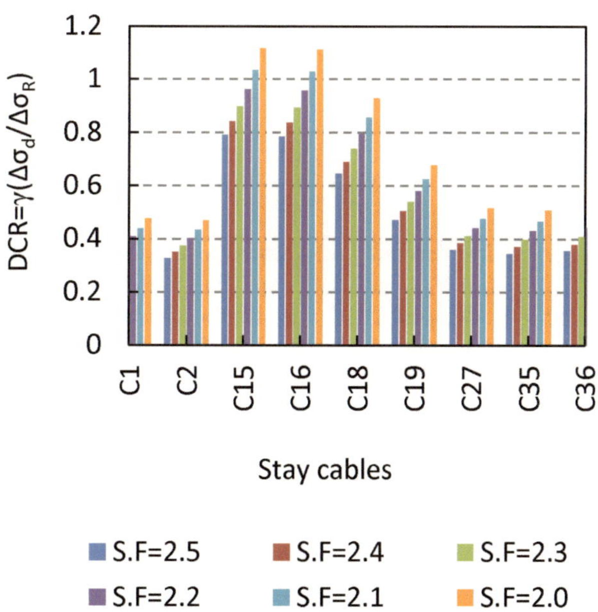

Figure 7.
Effect of fatigue load on DCR of stay cables of cable-stayed bridge.

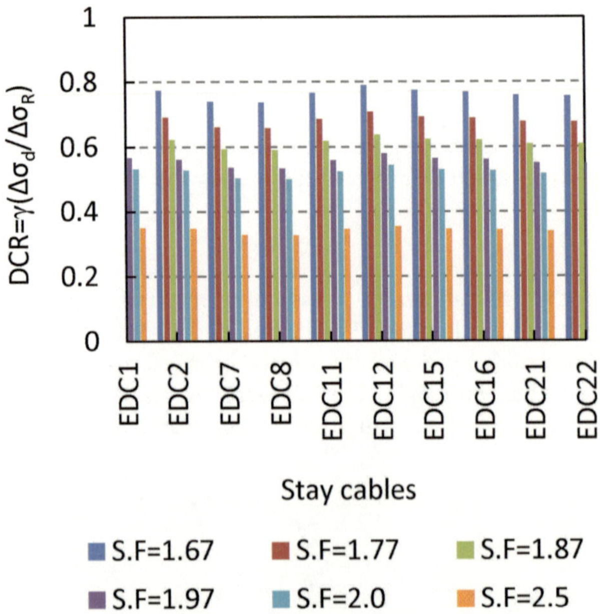

Figure 8.
Effect of fatigue load on DCR of stay cables of extradosed bridge.

shows that the safety factor of 1.67 satisfies the fatigue limit state. From probabilistic point of view, the safety of stay cables under the fatigue limit state is verified by satisfying the Palmgren-Miner hypothesis which states that fatigue failure of stay cables occurs when the accumulated damage exceeds one, $D(t) \geq 1$. Thus, if the fatigue failure time is denoted by T_f, then $P(T_f \leq t) = P(D(t) \geq 1)$. But this study is only limited to the deterministic fatigue analysis.

3.2 Ultimate limit state

After evaluation of safety factor of stay cables at fatigue limit state, the safety factor is further evaluated and verified at ultimate limit state. For that, following equation should be verified [9]:

$$\gamma_i \left(\frac{N_u}{N_{rd}}\right) \leq 1.0 \tag{6}$$

where γ_i is structural importance factor equal to 1.0, N_{rd} is equivalent design resistance of stay cables and N_u is ultimate axial load which is estimated by applying load and resistance factor design (LRFD) approach which considers the probabilities associated with simultaneous occurrence of different types of loads. Equations (7) and (8) yield ultimate axial loads for stay cables of cable-stayed and extradosed bridges, respectively [10]:

$$N_{u,CSB} = 1.25(DC + PS) + 1.5DW + 1.75(LL + IM) \tag{7}$$

$$N_{u,EDB} = 1.25DC + 1.5DW + PS + Pi + 1.75(LL + IM) \tag{8}$$

where the subscripts CSB and EDB are cable-stayed and extradosed bridges, respectively, DC is dead load (components and attachment), DW is dead load (wearing surface and utility), PS is pretension force, Pi is prestress force, LL is live load and IM is dynamic load allowance. In case of extradosed bridge, PS and Pi are not factored with the same coefficient of dead load. This approach is more reasonable for bridges with a rigid deck according to Mermigas [11].

In design viewpoint of long-span cable-supported bridges, PTI [12] suggests two methods. The first method consists of a simplified quasi-static analysis of cable-supported bridge with a missing cable under factored dead and live loads. These loads are combined with the static cable loss dynamic impact force (CLDF) resulting from the sudden breakage of a cable with the additional load factor of 1.1 on CLDF. In second method, PTI allows the usage of a dynamic analysis to compute the structural response more accurately due to an abrupt cable failure. However, little guidance is provided by PTI on how to conduct such a dynamic analysis. That is why, first method is selected in this paper for the sake of simplification.

The dynamic cable force is applied as an equivalent static force in the correct orientation on both anchorage points of cable by considering CLDF of 2.0 in the load combination. Following the aforementioned approach, the effects of cable loss on DCR of stay cables of cable-stayed and extradosed bridges are investigated thoroughly. **Figure 9** compares the DCR of stay cables of the cable-stayed bridge with and without sudden loss of single and multiple stay cables at different safety

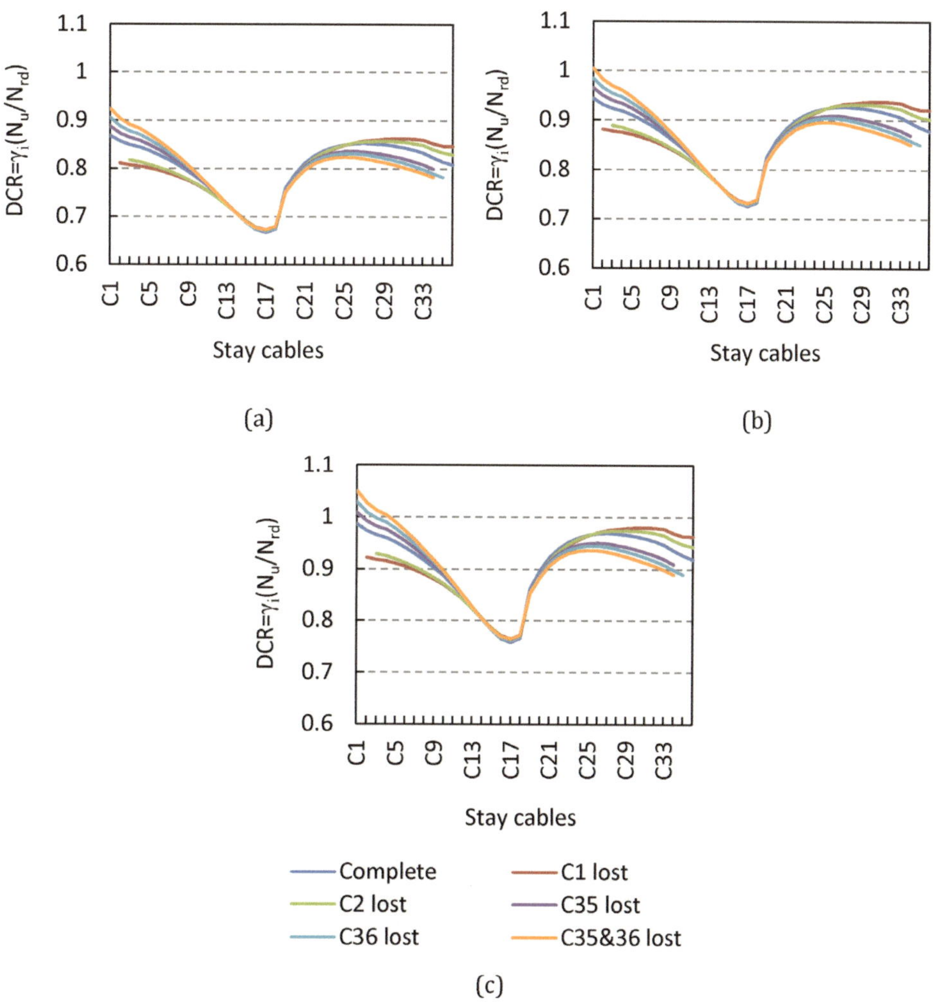

Figure 9.
Effect of cable loss on DCR of stay cables of cable-stayed bridge. (a) Safety factor of 2.5, (b) safety factor of 2.3, and (c) safety factor of 2.2.

factors. It can be observed from **Figure 9** that loss of two cables (C35&36) yields maximum DCR in the adjacent stay cables. This multiple cable loss event can also trigger the progressive collapse of the entire cable-stayed bridge.

Moreover, **Figure 9** also depicts that with the decrease of safety factor of stay cables, DCR increases accordingly and a minimum safety factor of 2.3 is essential to meet the requirements of ultimate limit state. Similarly, the effects of cable loss on DCR of EDCs are also investigated as shown in **Figure 10**. It is observed that the loss of two cables (EDC1&2) yields maximum DCR of EDCs and a safety factor of 1.67 is compulsory under normal loading condition which should be increased to achieve higher safety under extreme damaging condition.

In addition to that, the effect of corrosion as well as the combined effect of corrosion and cable loss on DCR of C1 and EDC1 are also examined at different safety factors in this study. For that, a simple corrosion model is adopted by introducing the uniform corrosion of 10% throughout the cable length as a change in cable area. The effective modulus of elasticity of corroded cable is determined and static analyses are performed. **Figure 11** shows that DCR of C1 is greater than 1.0 at a safety factor of 2.4 which indicates that the safety factor of 2.5 is the minimum factor required to avoid the rupture of C1. On the other hand, DCR of EDC1 is

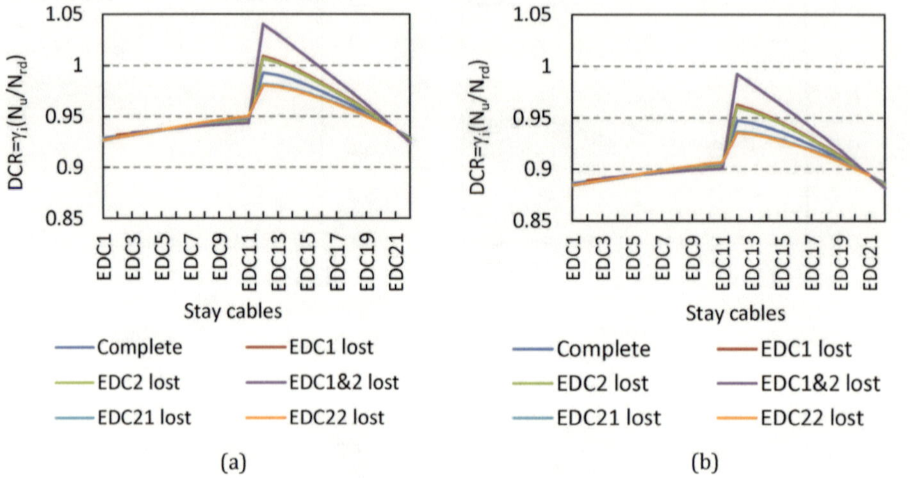

Figure 10.
Effect of cable loss on DCR of stay cables of extradosed bridge. (a) Safety factor of 1.67 and (b) safety factor of 1.75.

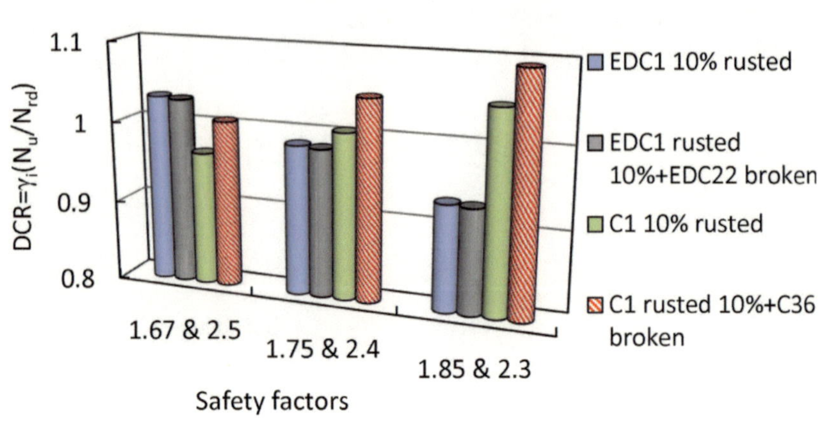

Figure 11.
Effect of corrosion and, combined effect of cable loss and corrosion on DCR of C1 and EDC1.

greater than 1.0 even at a safety factor of 1.67 which elucidates that a minimum safety factor of 1.75 is essential under extreme loading condition for the safe design of extradosed bridges.

4. Safety evaluation of stay cables by nondeterministic method

With the development of reliability-based methods, it has become evident that the traditional deterministic finite element method is not sufficient to properly design advanced structures or structural components subjected to a variety of complex loading conditions. Therefore, uncertainties in loads, material behavior and geometric configuration must be considered to provide rational reliability analysis and to describe the structural behavior with higher level of confidence.

In this paper, the safety factors of stay cables are also assessed by the nondeterministic method. For that, a probabilistic based reliability analysis code is prepared based on the mean value first order second moment (MVFOSM) reliability method. Basic random variables used for this program are material strength, dead loads and live loads. One million samples of normally distributed random variables are generated by using Monte Carlo simulation technique. The coefficient of variations (COV) of random variables are taken from the Ref. [13]. The program calculates the cable force (S) and resistance (R), and verifies the limit state function, i.e., $Z = R - S$ where R and S are linear and uncorrelated random variables. Subsequently, reliability index (β) and probability of failure (P_f) are determined from the relationships $\beta = \frac{\mu_Z}{\sigma_Z}$ and $P_f = \Phi(-\beta)$, respectively where μ_z is mean value, σ_z is standard deviation and Φ is cumulative distribution function for normal distribution.

For the acceptable values of probability of safety of structures, United States Army Corps of Engineers (USACE) suggests that the estimated reliability indices should be at least 3.0 (for above average performance) and 4.0 (for good performance) [14]. Based on it, the calculations of reliability index and failure probability for both bridge types are carried out and shown in **Tables 4** and 5. These tables clarify that reliability index decreases when safety factor decreases from 2.5 to 2.2 in case of cable-stayed bridge. For instance, the safety factors of 2.5, 2.3 and 2.2 yield

Safety factor	β	P_f
2.5	8.17	1.48×10^{-16}
2.4	6.79	5.31×10^{-12}
2.3	5.04	2.36×10^{-7}
2.2	2.91	1.8×10^{-3}

Table 4.
Reliability analysis results of C1 of cable-stayed bridge.

Safety factor	β	P_f
1.60	1.9	2.84×10^{-2}
1.67	4.37	6.03×10^{-6}
1.75	6.81	4.66×10^{-12}
1.85	9.32	5.76×10^{-21}

Table 5.
Reliability analysis results of EDC1 of extradosed bridge.

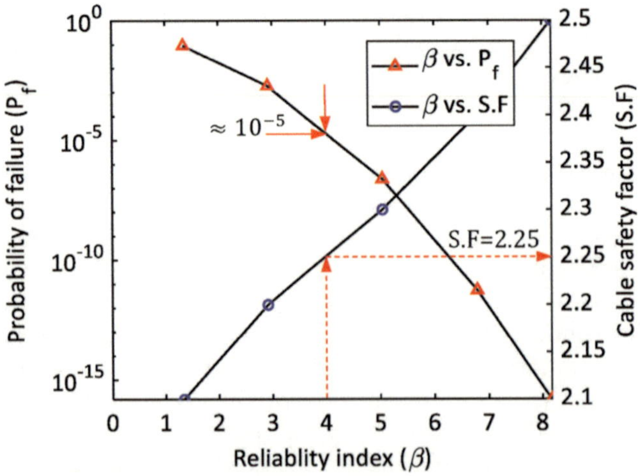

Figure 12.
Graphical evaluation of safety factor of C1 of cable-stayed bridge.

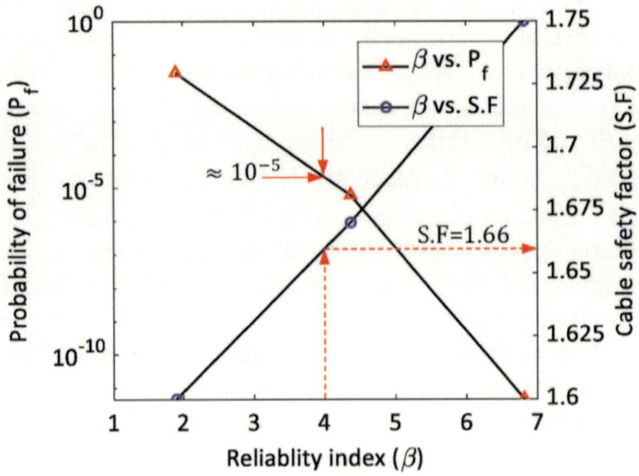

Figure 13.
Graphical evaluation of safety factor of EDC1 of extradosed bridge.

the reliability indices of 8.17, 5.04 and 2.91 for C1, respectively. Similarly, in case of extradosed bridge, the reliability index increases as safety factor increases from 1.60 to 1.85 for EDC1. The reliability analysis results also show that the safety factors of 2.3 and 1.67 yield the target reliability index greater than 4.0 for good performance of both bridge types. Based on these results, the optimum safety factors of C1 and EDC1 are calculated graphically as shown in **Figures 12** and **13**, respectively. It is observed that the safety factors of 2.25 and 1.66 yield the target reliability index of 4.0 and failure probability of 10^{-5} for stay cables C1 and EDC1, respectively. This also elucidates that the safety factor of 1.66 for extradosed bridges yields same reliability index as the safety factor of 2.25 for cable-stayed bridges.

5. Conclusions

In this paper, a parametric study on safety factor of stay cables of cable-stayed and extradosed bridges is carried out by using deterministic and nondeterministic methods. Following conclusions can be drawn from this study:

- Finite element analysis results show that cable-stayed and extradosed bridges are sufficiently redundant at safety factors ranging from 2.3 to 2.5 and 1.67, respectively under normal loading conditions. For cable-stayed bridges, ultimate strengths of stay cables are more critical than their fatigue strengths and a minimum safety factor of 2.3 is essential to satisfy the fatigue and ultimate limit states. However, in case of extradosed bridges, the ultimate strengths of stay cables are even more critical than their fatigue strengths and a minimum safety factor of 1.67 is indispensable to meet the limit state design requirements under normal loading conditions and it should be increased under extreme damaging conditions.

- The reliability analysis results elucidate that a minimum safety factor of 2.25 is necessary for stay cables of cable-stayed bridge to achieve the target reliability index of 4.0. Whereas, in case of extradosed bridge, a safety factor of 1.67 yields the reliability index greater than 4.0 and a minimum safety factor of 1.66 is essential for the safe design of extradosed bridges. Moreover, the safety factor of 1.66 for extradosed bridges yields same reliability index as the safety factor of 2.25 for cable-stayed bridges.

- The optimum safety factors evaluated by nondeterministic method are close to those obtained by deterministic finite element method. These outcomes imply that the structural reliability solutions for stay cables are rational and correct.

Author details

Khawaja Ali* and Aleena Saleem
Department of Civil Engineering, Yokohama National University, Japan

*Address all correspondence to: khawaja-ali-cd@ynu.jp

References

[1] Collings D, Gonzalez AS. Extradosed and cable-stayed bridges, exploring the boundaries. Proceedings of the Institution of Civil Engineers—Bridge Engineering. 2013;**166**(4):231-239

[2] Prestressed Concrete Technical Institute. Extradosed Bridge Design and Construction Standard. Japan: Gihodo Shuppan; 2009

[3] Ali K, Katsuchi H, Yamada H. Parametric study on cable safety of cable-stayed bridge considering ultimate and fatigue limit states. Journal of Structural Engineering, JSCE. 2018; **64**(A):99-108

[4] Japan Road Association. Japanese Specifications for Highway Bridges. Japan; 2002

[5] Elishakoff I. Safety factors and reliability: Deterministic actual stress & random yield stress. Safety Factors and Reliability, Friends or Foes. 2004:75-96. Available from: https://doi.org/10.1007/978-1-4020-2131-2_4

[6] Xiangyang W, Guanghui Z. Bridge reliability analysis based on the FEM and Monte-Carlo method. In: 2010 International Conference on Intelligent Computation Technology and Automation. 2010

[7] Zhang W, Cai CS. Fatigue reliability assessment for existing bridges considering vehicle speed and road surface conditions. Journal of Bridge Engineering. 2012;**17**(3):443-453

[8] Japanese Society of Steel Construction. Fatigue Design Recommendations for Steel Structures. Japan; 1995

[9] Japan Society of Civil Engineers. Standard Specifications for Steel and Composite Structures. Japan; 2007

[10] American Association of State Highway and Transportation Officials. LRFD Bridge Design Specifications. America; 2012

[11] Mermigas K. Behavior and design of extradosed bridges [thesis]. Toronto: University of Toronto; 2008

[12] Post-tensioning institute. Recommendations for Stay cable design, testing and installation. America; 2007

[13] Nowak AS. Calibration of LRFD Bridge Design Code. NCHRP Report 368. Washington, DC: Transportation Research Board, National Research Council; 1999

[14] Phoon KK, editor. Reliability-Based Design in Geotechnical Engineering: Computations and Applications. London and New York: Taylor & Francis; 2008